OWSO4R07e

ENERGY EFFICIENCY IN WASTEWATER TREATMENT IN NORTH AMERICA:

A COMPENDIUM OF BEST PRACTICES AND CASE STUDIES OF NOVEL APPROACHES

by:
George Crawford
Julian Sandino
CH2M HILL Canada Limited

2010

WERF

The Water Environment Research Foundation, a not-for-profit organization, funds and manages water quality research for its subscribers through a diverse public-private partnership between municipal utilities, corporations, academia, industry, and the federal government. WERF subscribers include municipal and regional water and wastewater utilities, industrial corporations, environmental engineering firms, and others that share a commitment to cost-effective water quality solutions. WERF is dedicated to advancing science and technology addressing water quality issues as they impact water resources, the atmosphere, the lands, and quality of life.

For more information, contact:
Water Environment Research Foundation
635 Slaters Lane, Suite G-110
Alexandria, VA 22314-1177
Tel: (571) 384-2100
Fax: (703) 299-0742
www.werf.org
werf@werf.org

This report was co-published by the following organization.

IWA Publishing
Alliance House, 12 Caxton Street
London SW1H 0QS, United Kingdom
Tel: +44 (0) 20 7654 5500
Fax: +44 (0) 20 7654 5555
www.iwapublishing.com
publications@iwap.co.uk

Library of Congress Catalog Card Number: 2010922833
IWAP ISBN: 978-1-84339-397-9/1-84339-397-2

CH2M HILL Canada Limited

The research on which this report is based was developed, in part, by the United States Environmental Protection Agency (EPA) through Cooperative Agreement No. CR83155901-2 with the Water Environment Research Foundation (WERF). However, the views expressed in this document are not necessarily those of the EPA and EPA does not endorse any products or commercial services mentioned in this publication. This report is a publication of WERF, not EPA. Funds awarded under the Cooperative Agreement cited above were not used for editorial services, reproduction, printing, or distribution.

ACKNOWLEDGMENTS

WERF thanks the volunteer efforts of the Optimization Challenge Interest Area Team, particularly the Technical Review Committee, for the support they gave to the preparation of this report. Research for this report was conducted under WERF contract OWSO4R07 Task Order 7.

The summary of existing best-practice documentation (presented in Chapter 2.0) was produced by HDR Inc., acting as a subcontractor to CH2M HILL, which is serving as the program manager for the WERF Optimization Challenge under Contract OWSO4R07 Task Order 7.

Research Team

Optimization Program Principal Investigator:
George Crawford
CH2M HILL Canada Limited

Task Order 7 Lead Investigator:
Julian Sandino, *CH2M HILL*

Project Team:
Ufuk Erdal
Todd Greely
Charles Shotts
Jay Surti
CH2M HILL

David J. Kinnear
David Reardon, *HDR Inc.*

Technical Review Committee

Joseph C. Cantwell, *Science Applications International Corporation (SAIC)*
Shahid Chaudhry, *California Energy Commission*
David Cooley, *Hampton Roads Sanitation District (HRSD)*
Robert F. Kelly, *Suez-Environnement*
David Tucker, *City of San Jose*
James Wheeler, *United States Environmental Protection Agency*
John Willis, *Brown and Caldwell*

WERF Optimization Challenge Issue Area Team

John Barber, *Eastman Chemical*
Shahid Chaudhry, *California Energy Commission*
Steve Constable, *DuPont Engineering Technology*
David Cooley, *Hampton Roads Sanitation District (HRSD)*
Robert F. Kelly, *Suez-Environnement*
Melanie S. Masek, *DuPont Engineering Research & Technology*

Arthur J. Meyers, Jr., *retired, University of Tennessee, adjunct professor*
Hsiao-Ting Ong, *City of San Jose*
Ali Oskouie, *Metropolitan Water Reclamation District of Greater Chicago (MWRDGC)*
Steven M. Rogowski, *Metro Wastewater Reclamation District of Denver*
Gary Shimp, *Black and Veatch*
Rob Simm, *Stantec Consulting, Inc.*
David Tucker, *City of San Jose*
Dariush Vosooghi, *City of Los Angeles*
Milind Wable, *CDM*
James Wheeler, *United States Environmental Protection Agency*
John Willis, *Brown and Caldwell*

Water Environment Research Foundation Staff

Director of Research: Daniel M. Woltering, Ph.D.
Program Director: Lauren Fillmore, M.Sc.

ABSTRACT AND BENEFITS

Abstract:

This report compiles North American best practices for the energy efficient operation of wastewater industry assets as a part the Global Water Research Coalition's (GWRC) project, *Energy Efficiency in the Water Industry: A Compendium of Best Practices and Case Studies,* which looks at these best practices worldwide. This report concisely presents the large volume of knowledge on well-established energy conservation and recovery best practices in wastewater treatment in North America. The report is a quick reference guide and comprehensive bibliography resource. It also documents case studies of novel energy conservation and recovery approaches/techniques and identifies implementation risks or obstacles, and management strategies.

Benefits:

- Serves as a starting point for wastewater treatment facilities wishing to implement energy conservation/recovery approaches and/or technologies, by providing details of implementation, including methodologies, techniques, strategies, and expected results.
- Identifies specific recommendations and anticipated outcomes related to improvements in energy efficiency through optimization of existing assets and operations through the implementation of well-established and documented best practices.
- Presents case studies of novel (yet proven at full scale) approaches with high potential of further improving energy conservation/recovery.

Keywords: Energy efficiency, energy management, energy conservation, energy recovery, wastewater treatment.

TABLE OF CONTENTS

LIST OF TABLES

LIST OF FIGURES

LIST OF ACRONYMS

AC	alternate current
AMD	acid manure digester
ARRA	*American Recovery and Reinvestment Act*
BOD	biochemical oxygen demand
CBFT3	Columbus Biosolids Flow-Through Thermophilic Treatment
CdTe	cadmium telluride
CH_4	methane
CHP	combined heat and power
CO_2	carbon dioxide
COD	chemical oxygen demand
DC	direct current
DO	dissolved oxygen
ECM	energy conservation measures
EPAct	*Energy Policy Act*
EPRI	Electric Power Research Institute
FOG	fats, oils, and grease
FP	focused pulsed
GHG	greenhouse gas
GWRC	Global Water Research Coalition
H_2	hydrogen gas
H_2S	hydrogen sulfide
IEEE	Institute of Electrical and Electronic Engineers
IEUA	Inland Empire Utilities Agency
mT CO_2 e/year	million metric tons of carbon dioxide equivalent per year
N	nitrogen
NACWA	National Association of Clean Water Agencies
NOx	nitrous oxide
NWWRP	Northwest Water Reclamation Plant
O&M	operations and maintenance
PD	positive displacement
PEF	pulsed electric field
PIER	public interest energy research
PTC	production tax credit
PV	photovoltaics
RP-1	Regional Plant 1
SAG	spherical artificial graphite

SAIC	Science Applications International Corporation
SCADA	supervisory control and data acquisition system
scf/day	standard cubic feet per day
SO	sulfur oxide
SRT	solids retention times
UASB	upflow anaerobic sludge blanket
USD	United States dollar
U.S. EPA	U. S. Environmental Protection Agency
UV	ultraviolet
VFD	variable frequency drives
VLM	vertical linear motion
WAS	waste-activated sludge
WEF	Water Environment Federation
WERF	Water Environment Research Foundation
WPCC	Water Pollution Control Center
WRP	water reclamation plant
WTF	water treatment facility
WWTF	wastewater treatment facility
WWTP	wastewater treatment plant

EXECUTIVE SUMMARY

Over the last decade, energy consumption by the water and wastewater sector has considerably increased as a result of implementation of technologies to meet new effluent and potable water quality standards. The price of energy has also substantially increased in the same period. Some North American and European water utilities have reported increases in energy costs of over 60% in recent years; with oil prices continuing to fluctuate; further substantial increases in operating costs are possible. Potential cost increases will be compounded by the need to meet additional new regulations that will require energy intensive treatment processes to achieve tight standards. High energy consumption will affect the water industry worldwide and is inextricably linked to the issue of climate change.

To address these issues, the Global Water Research Coalition (GWRC) has undertaken development of a compendium of best practices in the energy efficient design and operation of water industry assets worldwide. As the North America practice coordinator for this GWRC effort, the Water Environment Research Foundation (WERF), has identified specific recommendations and anticipated outcomes related to improvements in energy efficiency through optimization of existing assets and operations through the implementation of well-established and documented best practices and presented case studies of novel (but proven at full scale) approaches with high potential of further improving energy conservation/recovery.

The information developed under this research project is intended to serve as a starting point for wastewater treatment facilities wishing to implement energy conservation/recovery approaches and/or technologies, by providing details of implementation, including methodologies, techniques, strategies, and expected results.

Best-in-class wastewater utilities strive to continuously improve efficiency in all aspects of operations, particularly energy optimization. Therefore, a history of successful municipal wastewater treatment energy conservation measures exists over the past 20 years in North America. Chapter 2.0 summarizes best practices in energy optimization/energy recovery technologies and/or practices, as developed from the review of documentation recently published by an assortment of private and public entities. This information is intended to be used by municipal wastewater treatment facilities in North America interested in implementing an energy optimization program. The report documents information related to wastewater treatment facility energy consumption patterns, aspects related to energy economics (e.g., power utility rate structures, peak-shaving strategies, and life-cycle cost assessment methodologies), as well as presents summaries of well-established and documented energy conservation measures such as in-plant energy management approaches, and the incorporation of technology advancements for energy recovery and the optimization of energy-intensive processes such as pumping and aeration.

Chapter 3.0 presents case studies of Energy Optimization/Energy Recovery Technologies and/or Practices information from novel, full-scale case studies that were obtained from a variety of means, including engineering reports, journal articles, interviews and facility visits. In general terms, the information included provides a general overview of the technology/practice (independent of the specific case study), and for a particular facility, it establishes the background, the type and size of the process, the situation before and after any changes, the changes themselves and the results obtained. The information collected is based on verifiable, full-scale sources, so that results could be quantified, allowing for the identification of implementation risks or obstacles, and means by which they were managed. Following is the list of the 13 novel technologies/approaches case studies documented:

- Advanced Anaerobic Digestion and Combined Heat and Power (CHP) – Columbus Biosolids Flow Through Thermophilic Treatment (CBFT3) Technology, South Columbus Water Resource Facility, Columbus, Georgia.
- Co-Digestion of Dairy Manure with WWTF Sludge – Inland Empire Utilities Agency Regional Plant 1, City of Ontario, California
- Sludge Reduction Technologies – Focused Electrical Pulse, OpenCEL FP Unit at the Northwest Water Reclamation Plant, Mesa, Arizona.
- Biogas Cleaning Technologies – Siloxane Removal System, Barrie Water Pollution Control Centre (WPCC), Ontario, Canada.
- Anaerobic Digester Mixing – Linear Motion Mixers, Ina Road Wastewater Treatment Plant (WWTP), Tucson, Arizona
- Co-Generation – External Combustion Engines for Anaerobic Digestion Biogas, Corvallis Wastewater Reclamation Facility (WWRF), Corvallis, Oregon.
- Fuel Cells Operated with Anaerobic Digester Gas – South Treatment Plant, King County, Washington.
- Microturbines Operated with Anaerobic Digester Gas – Lancaster Water Reclamation Plant (WRP), Lancaster, California.
- Wind Power – Jersey – Atlantic Wind Farm, Atlantic County Utilities Authority Wastewater Treatment Facility (WWTF), Atlantic City, New Jersey.
- Solar Power – Inland Empire Utilities Agency Carbon Canyon Water Reclamation Facility (WRF), City of Chino, California.
- Hydropower – City of San Diego Point Loma WWTP Hydroelectric Generation System, Point Loma, California.
- Anaerobic Treatment of Municipal Wastewater with Upflow Anaerobic Sludge Blanket (UASB) Reactors – Rio Frio WWTF, Bucaramanga, Colombia.

CHAPTER 1.0

INTRODUCTION

High energy consumption affects the wastewater industry worldwide, and is second in cost only to manpower for most wastewater utilities. Over the last decade, the implementation of new technologies to meet new effluent limits and water quality standards has considerably increased energy consumption, and the price of energy has also substantially increased. In North America and Europe, some utilities have reported significant increases in energy costs in recent years, and with oil prices continuing to fluctuate, further substantial increases in operating costs could be expected. Those increases will be compounded by the need to meet additional new regulations that will require energy-intensive treatment processes to achieve tight standards. High energy consumption will affect the wastewater industry worldwide and is inextricably linked to the issue of climate change.

Through its Optimization Challenge Program, the Water Environment Research Foundation (WERF) is serving the role of North America wastewater practice coordinator in the Global Water Research Coalition's (GWRC) project titled *Energy Efficiency in the Water Industry: A Compendium of Best Practices and Case Studies*. Through this assignment, WERF intends to define specific recommendations regarding:

- Incremental improvements in energy efficiency through optimization of existing assets and operations
- More substantial improvements in energy efficiency from the adoption of novel (but proven at full-scale) technologies

As part of the GWRC project, WERF researchers summarize existing information on well-established energy optimization/energy recovery best practices, and documents a series of case studies of novel (yet full-scale proven) technologies/practices in wastewater treatment primarily in North America.

The report documents the case studies of energy optimization/energy recovery technologies and/or practices considered being novel and full-scale proven in at least one installation. Information from these case studies was obtained from a variety of means, including engineering reports, journal articles, interviews, and facility visits.

CHAPTER 2.0

SUMMARY OF BEST PRACTICES IN ENERGY OPTIMIZATION/ENERGY RECOVERY TECHNOLOGIES AND/OR PRACTICES

2.1 Introduction

Wastewater utilities must produce effluent that conforms to regulatory requirements as efficiently as possible to keep operating budgets and, therefore, user charges aligned with public expectation. Energy costs greatly influence utility operating budgets, with only debt service or labor contributing a greater fraction to the overall budget.

Best-in-class wastewater utilities strive to continuously improve efficiency in all aspects of operations, particularly energy optimization. Therefore, a history of successful municipal wastewater treatment energy conservation measures (ECM) exists over the past 20 years in North America. This document summarizes well-established, proven ECM and presents a review of energy-optimization-related documentation for use by municipal wastewater treatment facilities (WWTFs) interested in implementing an energy optimization program.

2.2 Wastewater Treatment Facility Energy Consumption

The 16,000 publically owned U.S. WWTFs consume significant quantities of electrical energy, estimated to be approximately between 1-4% of total energy production, varying regionally or approximately 40 million megawatts per year (MWh/yr). At the average U.S. electrical price (September 2009) of USD$7.18 cents/kilowatt-hours (kWh), this amounts to USD$2.8 billion being spent on electrical power for wastewater treatment country-wide in 2009.

As an example, Table 2-1 presents data summarized in a recent *Energy Policy Act* (EPAct) energy management guidebook (U.S. EPA, 2008) from the State of Massachusetts, which agrees well with the entire U.S. in terms of unit energy production, although the average electricity price of USD$12.9 cents/kWh falls on the high side.

Table 2-1. Massachusetts Annual Wastewater Energy Use Summary.

Unit Electricity Use	1,750	kWh/mg
Total Energy Consumption	707,735	MWh/yr
Total Energy Cost	USD$91.3	million dollars

A recent Water Environment Federation (WEF) water and wastewater treatment facility energy conservation guidebook (WEF, 2009) provides detailed energy consumption benchmarks for different types of WWTFs. Table 2-2 presents the average unit total electrical consumption from these facilities.

Table 2-2. Unit Electrical Consumption, kWh/d.

	4-ML/d	20-ML/d	40-ML/d	75-ML/d	190-ML/d	380 ML/d
	1-mgd	5-mgd	10-mgd	20-mgd	50-mgd	100-mgd
Trickling Filter	1,811	978	852	750	687	673
Activated Sludge	2,236	1,369	1,203	1,114	1,051	1,028
Advanced without Nitrification	2,596	1,573	1,408	1,303	1,216	1,188
Advanced with Nitrification	2,951	1,926	1,791	1,676	1,588	1,558

The table provides an initial benchmark of WWTF unit energy consumption, demonstrating the trend that unit power increases as effluent requirements become more stringent, and decreases asymptotically as facility size increases. Although benchmarks do provide an indication of how facilities compare, many physical factors influence facility energy consumption and need to be considered in a more detailed analysis. For example, pumping requirements depend on the topography of the facility site, and diffused aeration efficiency depends on the aeration basin depth, factors which cannot be changed at a reasonable cost. Some facilities can improve significantly beyond the benchmark, while others can be fully optimized and still not meet the benchmarks.

Energy consumption varies considerably between wastewater unit processes and between different wastewater facilities, but several trends exist. Figure 2-1 on the following page presents typical wastewater facility energy consumption. The figure indicates that aeration consumes more than half of the electricity in this typical WWTF, which does not recover energy from anaerobic digestion biogas production and applies chlorination rather than ultraviolet (UV) disinfection. Pumping consumes approximately 14% of the energy, as does anaerobic digestion mixing. Depending on the facility topography, pumping costs as a percentage of the overall energy costs can vary considerably. Anaerobic digestion, as discussed in this report, can recover approximately 30 to 40% of the overall energy consumption through CHP energy recovery systems. UV disinfection can increase energy requirements between 7% and 15%.

Although the USD$2.8-billion wastewater electrical bill seems like a very large number, it is being shared by 231 million U.S. citizens serviced by sewer systems for a cost of approximately USD$12 per citizen per year. Wallis-Lage et al. (2009) provides a discussion of this. In the USD$13.5-trillion economy, this represents just 0.02% of economic activity. These numbers indicate that the wastewater community delivers reasonable and efficient service.

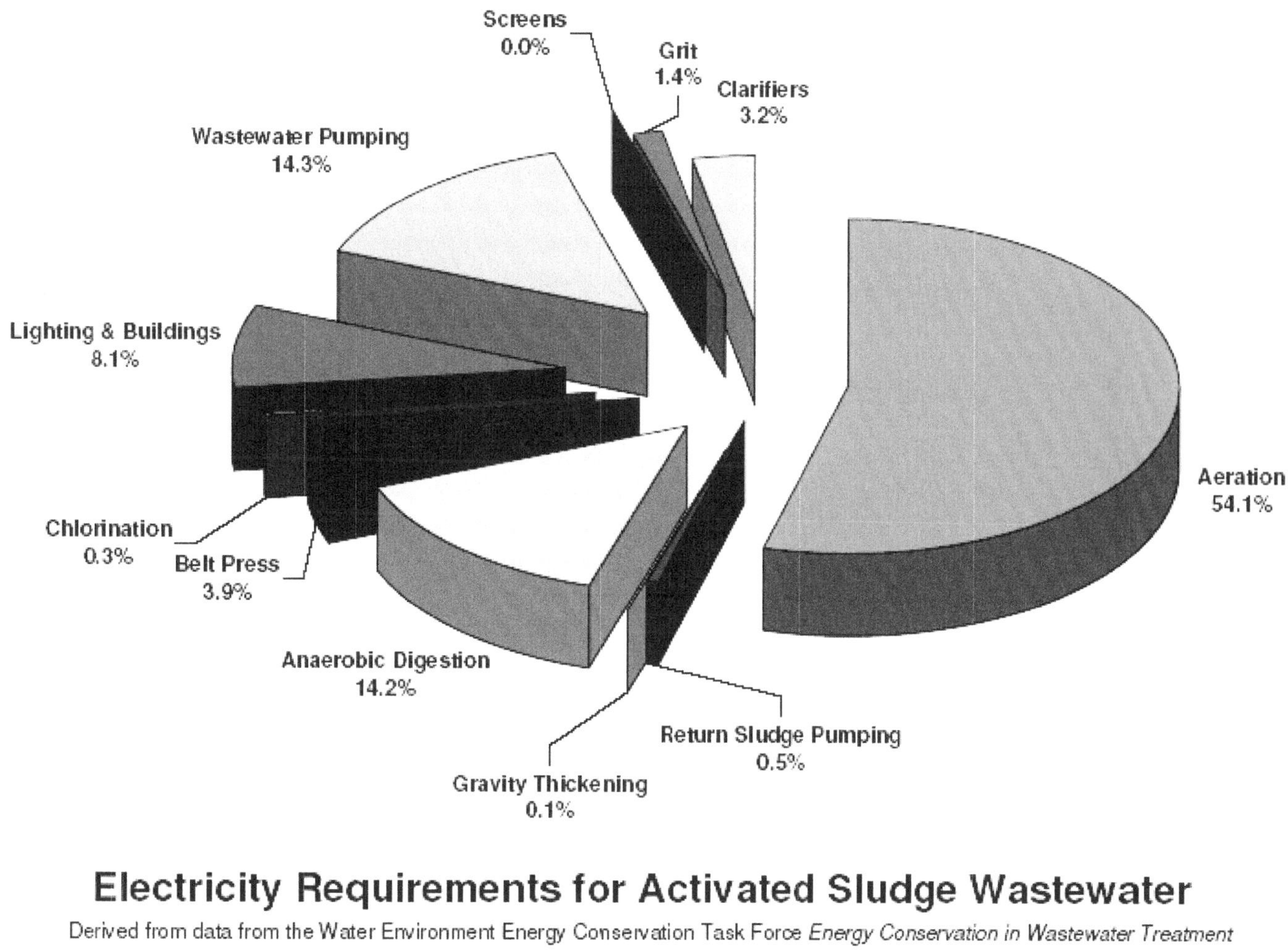

Figure 2-1. Example Electrical Requirements for Activated Sludge Wastewater Treatment.
Source: (SAIC, 2006).

Many of the documents reviewed and discussed in this report emphasize the urgency of implementing ECM to respond to increasing energy prices. Figure 2-2 presents both nominal and real U.S. average electricity prices (source: Energy Information Administration) and National Association of Clean Water Agencies (NACWA) Sewer Rates (NACWA, 2008) between 1996 and 2007.

Figure 2-2 indicates that although electricity prices increased in nominal terms, they remained constant and even decreased in real terms. Over the entire period, electricity increased only 0.7% in real terms. Sewer rates increased by 11% in real terms during the period for which the consumer price index increased 32%. Although electricity prices did not keep pace with general inflation, neither did sewer rates. The 2008 NACWA Financial Survey indicates that per capita wastewater utility total expenditures increased 47% between 1999 and 2007. Significant portions of wastewater utility budgets, particularly debt service and chemical costs, could continue to outpace inflation in the upcoming years. Both electricity and labor can be controlled to some degree, but management must optimize cautiously while still providing a level of service to accomplish the utilities' missions without compromising important requirements such as water quality or employee safety.

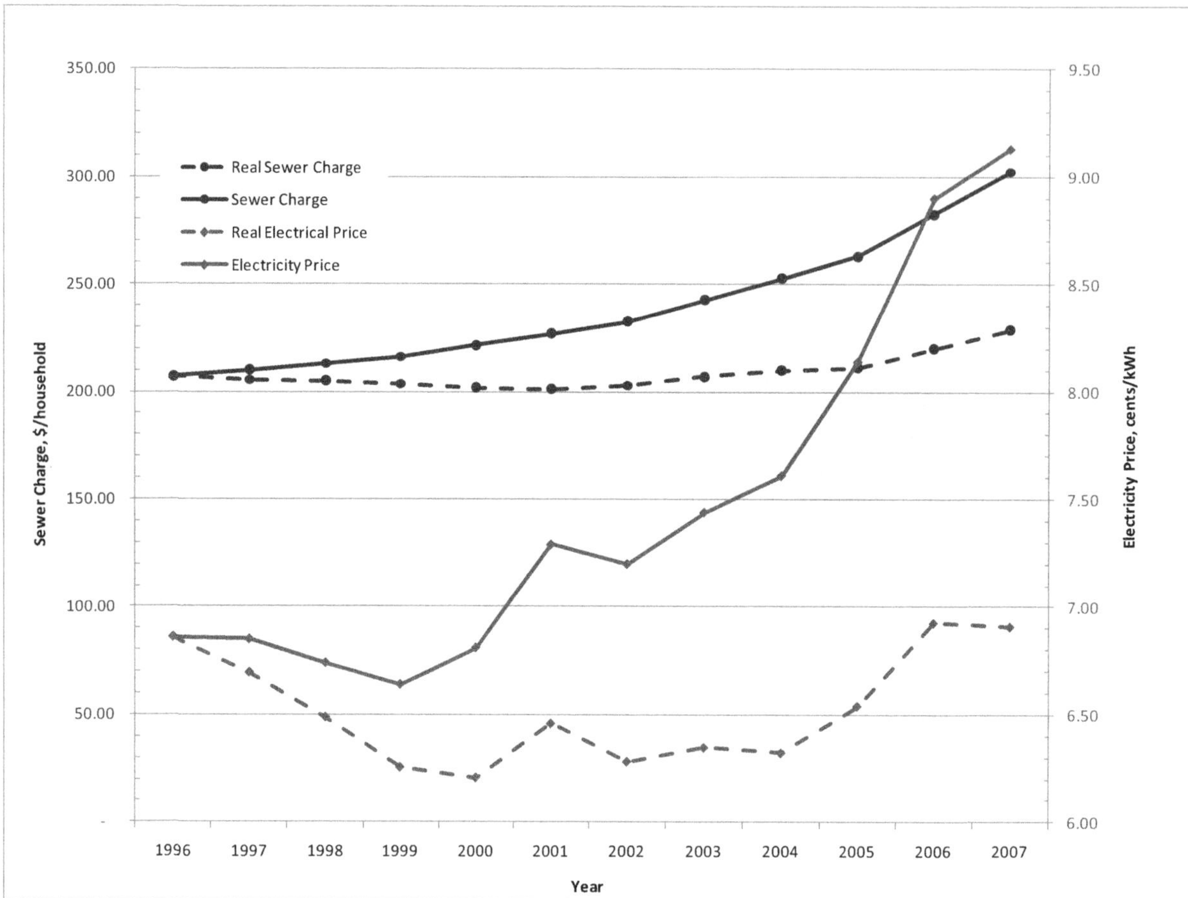

Figure 2-2. U.S. Average Electricity Prices and Sewer Rates (1996–2007).

In the past 10 years, climate change concerns have provided another motivation for energy optimization, because emission of carbon dioxide (CO_2) due to power consumption makes up the largest fraction – as much as 80% – of a WWTF greenhouse gas (GHG) footprint. Excessive anthropogenic GHG emissions increase global temperatures, creating climate change and concerns about the impact on earth's ecosystems.

Many GHG regulations, including the United States Environmental Protection Agency (U.S. EPA) mandatory reporting of GHG promulgated on October 30, 2009 (U.S. EPA, 2009) differentiate three emission scopes to avoid double counting GHGs from related sources: Many carbon footprint calculations (The Climate Registry, International Council for Local Environmental Initiatives, etc.) differentiate three scopes to avoid double counting GHG emissions from related sources.:

- Scope 1 – Direct GHG emissions
  - Production of electricity, heat, or steam
  - Physical or chemical processing
  - Transportation of materials, products, waste, and employees
  - Fugitive emissions
- Scope 2 – GHG emissions from imports of electricity, heat, or steam

- Scope 3 – Other indirect GHG emissions: Consequences of the activities of the reporting company but occurring from sources owned or controlled by another company
  - Employee business travel
  - Transportation of products, materials, and waste
  - Outsources activities
  - Emissions from waste
  - Emissions from final product disposal
  - Employee commuting
  - Production of imported materials

The U.S. EPA mandatory reporting regulation promulgated on October 30, 2009, requires facilities emitting greater than 25,000 million metric tons of carbon dioxide equivalent per year (mt CO_2 e/yr) Scope 1 GHG emissions to report to U.S. EPA annually. Municipal WWTFs are expected to be below this reporting threshold.

Municipal WWTFs should set an example to the communities they serve and reduce GHG emission as much as possible, and energy conservation provides one of the best opportunities. Combustion of biogas generated through anaerobic digestion in CHP systems, as discussed here, offers a method for municipal WWTFs to reduce GHG emissions by offsetting fossil fuel combustion. For both economic and environmental reasons, it should also be explored.

2.3 Energy Economics

2.3.1 Power Utility Rate Structures

In the U.S., electrical utilities traditionally function as a regulated monopoly, providing electrical service exclusively in a region with rates approved by a state public utility commission and the Federal Energy Regulatory Commission. To provide a more competitive marketplace as permitted under the EPAct of 1992, different regions of the United States considered, and some proceeded with, deregulating electrical utilities. Deregulation allows customers to choose an electrical service provider from among several energy companies serving the same region, but would remove public utility commission approval on rate structures. Economically similar to phone service deregulation, electrical utility deregulation could improve efficiencies and reduce cost to the consumer because of increased competition, but would also allow electrical providers to charge whatever the market bears. Troublesome deregulation experiences in California, technical geographic complications, and concern over safety, particularly in nuclear facilities, compel deregulation to proceed cautiously.

Understanding the utility bill in some ways represents the first step in optimizing energy efficiency for a WWTF. As discussed below, peculiarities of utility rate structures mean that minimizing energy usage does not always mean minimizing cost, something WWTF management needs to consider.

Typical WWTF utility bill charges can include the following categories:

- Customer Charge
- Energy Charge
- Demand Charge
- Power Factor Surcharge

- ♦ Fuel-Cost Adjustment
- ♦ Regulatory Fees
- ♦ State and Local Taxes
- ♦ Transmission Voltage
- ♦ Standby Service

This section discusses each of these categories individually.

The Customer Charge compensates a utility for administrative costs incurred in servicing the customer. Compared to other charges, the customer charge typically does not represent a large percentage of the overall bill.

The Energy Charge compensates the energy utility for operating costs for generating and supplying electricity, including profit. Billing depends on energy usage as determined by the meter reading in kilowatt-hours (kWh) during the billing period.

Although billing depends on energy used, utilities may not charge the same rate per kWh during all periods. Billing rates may vary both diurnally and seasonally. For example, Table 2-3 presents an Energy Charge Rate Schedule for a typical WWTF billing structure, and Table 2-4 presents an Energy Charge Peaking Schedule for a WWTF. The rate charged in this example depends on both diurnal and seasonal periods to reflect those periods, particularly in the summer, when the utility must strain resources to provide service.

Table 2-3. Typical Energy Charge Rate Schedule.

Peak Condition	Summer Months June – September cents/kWh	Winter Months October – May cents/kWh
All On-Peak Energy per Month	8.2512	7.4545
All Off-Peak Energy per Month	5.5573	5.2389

Table 2-4. Typical Energy Charge Peaking Schedule.

Peak Condition	Summer Months June – September	Winter Months October - May
On-Peak Period Hours	1:00 p.m. – 9:00 pm Monday – Friday	6:00 a.m. – 1:00 p.m. Monday – Friday
Off-Peak Period Hours	All other weekday hours and all Saturday and Sunday hours. All hours for following holidays: New Years Day, Memorial Day, Good Friday, Independence Day, Labor Day, Thanksgiving and Following Day, and Christmas Day	

The Demand Charge depends on the maximum power drawn during the billing period (typical averaged over a contiguous 15-minute or longer period) as the combined power of all demand served by the main meter. Demand charges allow electrical utilities to recover the capital or fixed costs of providing power, including debt service for power plants, transmission lines, transformers, right-of-ways, etc. These fixed costs relate in a complex manner to the maximum demand charge and therefore serve as a surrogate to allocate these costs among the user base in a reasonable manner. Table 2-5 presents an example Demand Rate Schedule for a typical WWTF. Most sizable WWTFs would demand a maximum of greater than 5,000 kilowatts (kW) in a month and therefore would apply the maximum rate on this schedule.

Table 2-5. Typical Demand Rate Schedule.

Total Billing Demand per Month	Summer Months June – September USD$/kW	Winter Months October – May USD$/kW
First 2,000 kW of Billing Demand	15.98	10.64
Next 3,000 kW of Billing Demand	13.89	8.54
Over 5,000 kW of Billing Demand	11.79	6.43

A Power Factor Adjustment or surcharge may apply if a WWTF power factor falls below a certain value, often 0.85 because utilities recover costs for the actual power but must provide facilities to provide the apparent power. To maintain a power factor close to 1.0 and avoid power factor surcharges, WWTFs should employ the strategies discussed later in this document.

Fuel cost adjustments may apply to compensate electrical utilities for fuel price volatility, which keeps published energy charges more constant while the fuel surcharge varies. Regulatory fees may apply to satisfy cost deficiencies resulting from fuel-cost apportionment or as mandated by a legislature. Publically owned WWTFs should be exempt from State and Local taxes and therefore should not be charged. Transmission voltage charges may apply or discounts may apply if a WWTF provides transformers to step voltages down from transmission voltages to applied voltages within the WWTF. Finally, standby service charges or discounts may apply when a WWTF requires multiple independent power sources as standby service. Service charges may apply when the electrical utility provides standby. Discounts may apply if a WWTF provides its own standby service with emergency generators. In these cases, a discount can be negotiated that provides the WWTF with a price incentive to decrease demand by using emergency generators at the discretion of the electrical utility.

2.3.2 Peak-Shaving

Utility rate structures, as presented in Tables 2-3 through 2-5, offer opportunities to minimize energy cost by moving demand into periods with lower cost or avoiding high-cost energy rates altogether, decreasing the energy cost. Two strategies can be employed:

- Flow Equalization/Peak Storage – WWTFs with adequate equalization volume may elect to treat wastewater during off-peak billing hours by storing flow during on-peak and treating during off-peak. The associated energy consumption, particularly aeration, therefore occurs at a lower rate. The capital cost of equalization, including basin construction, some operating costs, and potentially pumping, must be evaluated in the life-cycle. In some cases, equalization with capital costs required for peak equalization can be more effectively utilized though this technique, particularly if hydraulics are favorable.

- Emergency Generator Operation – Emergency generators required for redundancy can be employed to decrease energy costs through two strategies:
  - Reducing the demand charge by operating if the peak demand exceeds a targeted value. Supervisory control and data acquisition systems (SCADA) monitoring of electrical usage and automatic emergency generator operation would be required to implement this strategy.
  - Reducing the energy charge by operating when the marginal cost of generator operation exceeds the marginal energy cost of service provided by the electrical utility. WWTFs employing this strategy must consider GHG emissions, because they, in essence, become a power generator increasing GHG emissions. Operating in a manner that exceeds the U.S. EPA Mandatory Reporting of GHGs threshold may not be a sound strategy.

These strategies do not necessarily minimize the energy consumption; however, and WWTF management should consider the consequences and develop a plan. Should a WWTF attempt to minimize energy consumption or energy cost? Probably a mixed strategy that considers both should be employed, taking into consideration the energy production mix by the electrical utility and GHG emission factors. For example, a WWTF served predominantly by hydropower may elect to maximize utility usage and minimize diesel-powered emergency generator usage, even at a higher cost for carbon footprint reasons. The regional and national electrical grid complicates this policy decision.

2.3.3 Life-Cycle Cost Assessment

The complex nature of electrical utility billing structures, the variable demand of a WWTF, and many other factors complicate energy conservation decision-making. The GHG example in the previous section demonstrates that multiple objectives need to be weighed frequently in deciding both initial design concepts and which energy conservation measures should be implemented.

Regardless of the cost analysis being considered, the life-cycle of ECM should be considered in the economic decision process. For many pieces of high-energy consumption equipment, the operating costs can be many times greater than the capital costs. Therefore, the present worth or annual cost of a project should be developed, amortizing capital costs and operating costs consistently.

2.4 North American Energy Conservation Measures

This section summarizes proven North American ECMs and serves as a starting point for facilities wishing to implement them. The references provide details of implementation, including methodologies, techniques, strategies, and expected results. Several states have mature energy efficiency programs oriented to the wastewater sector (California Energy Commission, New York State Energy Development Authority, and Wisconsin Focus on Energy). Based on the experience of these programs, data on implementing energy efficiency measures at multiple WWTFs is presented in the following sections.

2.4.1 Management Techniques

Successful ECM implementation requires management commitment to demonstrate the importance of energy conservation to the entire WWTF staff. U.S. EPA (U.S. EPA, 2008) provides excellent resources for management techniques useful in energy optimization projects considering the following programs:

- Energy Star
- Asset Management
- ISO 14001 Environmental Management Systems
- American National Standards Institute/Management System for Energy 2000: A Management System for Energy

The U.S. EPA recommends that management take a seven-step approach to energy conservation and demonstrates how each of the programs incorporates the following steps:

1. Getting Ready – Make a commitment
2. Assessing Current Energy Baseline – Determine the current WWTF energy usage and benchmark this to similar facilities in terms of energy usage and cost for both the entire facility and for each of the major power demands in the WWTF
3. Energy Vision and Priorities for Improvement – Institute an energy policy that defines the motivations for energy efficiency, including cost and environmental impacts
4. Objectives and Targets – Define energy usage targets clearly
5. Energy Improvement Management Plan – Develop a plan on how to achieve the defined targets, including specific metrics, actions, and a compliance schedule
6. Monitoring and Measuring – Evaluate progress by measuring and charting energy consumption
7. Maintaining Energy Improvements Program – Recognize and publicize the achievements to encourage continuous improvement and demonstrate management commitment (unique to the U.S. EPA approach and Energy Star)

2.4.2 Lighting Systems

Advances in lighting systems provide the WWTF with opportunities to reduce energy consumption by retrofitting or providing lighting that is more efficient. Energy Power Research Institute (EPRI, 1998) provides a detailed discussion of possible lighting retrofits, including the following:

- Fluorescent lighting upgrades such as compact fluorescent lamps
- Light-emitting diode array upgrades
- Fixture upgrades
- Incandescent lighting upgrades
- High-intensity discharge lighting upgrades
- Occupancy sensor installation
- Scheduling controls installation

Lighting upgrades provide ancillary benefits beyond reduced energy costs, including maintaining or improving lighting quality, reducing maintenance costs, and reducing heat output. Energy consumption reduction and cost-effectiveness vary considerably depending on the application and existing technology, with energy cost reductions of 25% to 75% reported.

2.4.3 Liquid Treatment Process Selection, Operation, and Power Requirements

Designers must select appropriate treatment processes to meet or exceed effluent requirements but must also be aware that different processes consume different quantities of power Figure 2-3 provides general guidance on the power requirements of different treatment processes for facilities greater than 1 mgd.

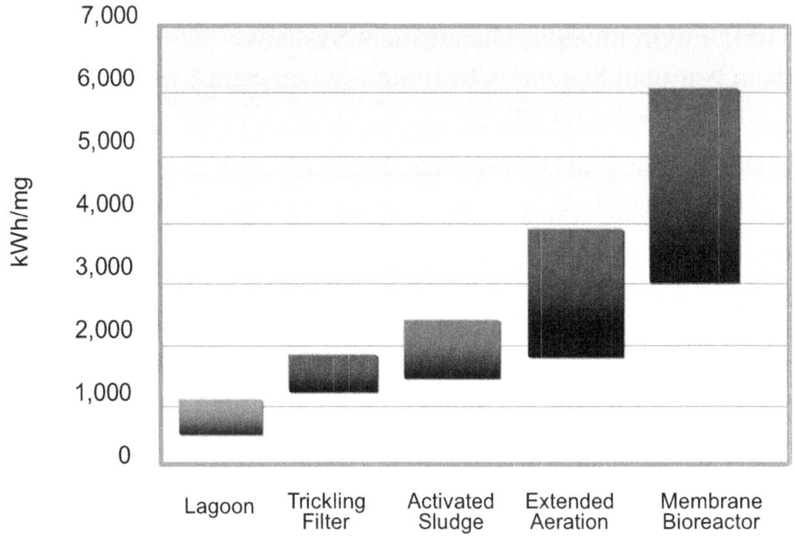

Figure 2-3. Treatment Process Power Requirements.

Intuitively, increased levels of treatment require increased quantities of power. Designers and decision-makers need to plan for possible stringent future effluent requirements; at the same time, they need to consider optimizing operational costs. Balancing these two opposing interests can be a challenge.

The addition of a pre-anoxic zone and internal mixed-liquor recycle to a nitrifying activating sludge system in a modified Ludzack-Ettinger configuration reduces the energy costs by using readily-biodegradable chemical oxygen demand (COD) as a carbon source for denitrification. This COD, therefore, does not require that dissolved oxygen (DO) be provided for its respiration by heterotrophic microorganisms. Estimates of the relative costs of conventional treatment, nitrification, and nitrification/denitrification are presented in Figure 2-4 (Rosso, 2007). The figure indicates nitrification increases aeration energy by 33% compared to non-nitrifying conventional activated sludge (assigned a reference factor of 1.0), but nitrification/denitrification reduces costs to 88% of the reference value.

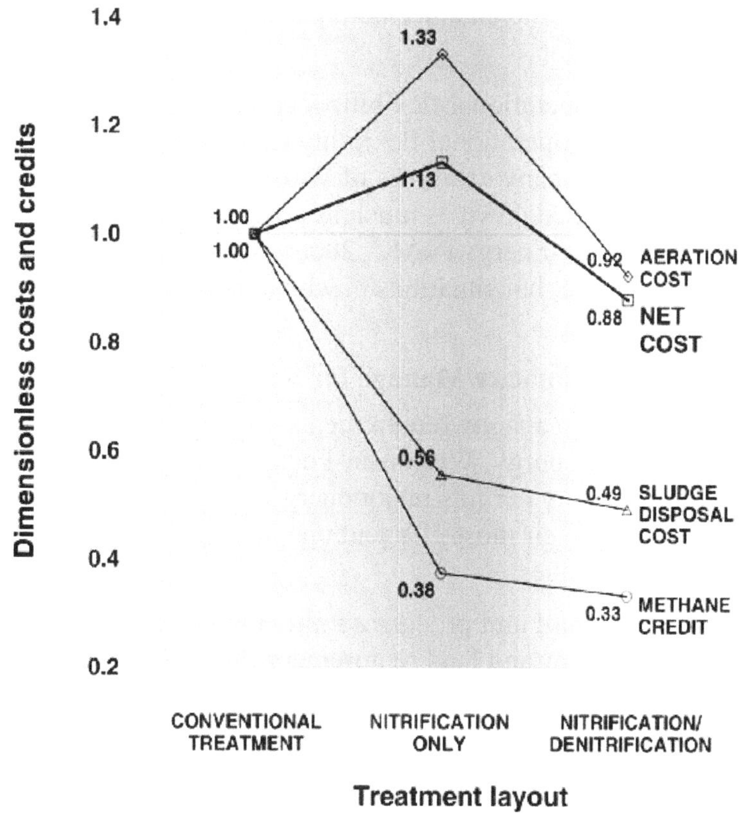

Figure 2-4. Relative Costs of Nitrification/Denitrification Treatment Processes.

Reprinted with permission from Rosso, 2007.

2.4.3.1 Operational Flexibility

Designing WWTFs for operational flexibility, as described by the Water and Wastewater Energy Best Practice Guidebook [Science Applications International Corporation (SAIC), 2006], allows operators more opportunity to conserve energy. Wastewater equipment that consumes a high level of energy, such as pumps and blowers, operates more efficiently at its design point. Providing flexibility offers opportunities such as the following, which can save energy:

1. Taking unit processes like aeration tanks out of service during periods of low flow, low organic loading, and/or high temperature
2. Providing variable-speed drives on appropriate equipment, to match duty to performance requirements
3. Providing several smaller pieces of equipment, such as blowers, for service rather than a fewer number of larger pieces of equipment to allow better matching of duty requirements for diurnal, seasonal, and growth variation
4. Installing reliable and appropriate instrumentation to provide operators with rapid, direct feedback on operational requirements such as dissolved oxygen and nutrient probes. In many cases, these instruments can be integrated into control loops to provide energy-saving automatic operation.
5. Providing "swing" bioreactors designed for operation in several metabolic zones, such as aerobic, anoxic, or anaerobic. This allows optimization of the hydraulic retention times in

the various metabolic zones to enhance both performance and energy use as organic load changes over time or season.

The benefits of built-in operational flexibility can be difficult to estimate because a WWTF that is well designed for operational flexibility cannot be easily compared to one that has been poorly designed. There are many examples of WWTFs throughout the industry whose blowers are in operation and turned down as much as possible but are still bleeding air to the atmosphere. Wisconsin-Focus-on-Energy (SAIC, 2006) estimates that 10% of 25% of energy can be conserved with flexible design, but situations in which more can be saved could be possible in extreme situations.

2.4.3.2 Staging of Treatment Capacity/Manage for Seasonal and Tourist Peaks

Forecasting and planning capacity requirements, as well as considering seasonal and tourist variability, can conserve energy. Wisconsin-Focus-on-Energy (SAIC, 2006) describes how the benefit of staging capacity permits major energy-consuming equipment, such as blowers and pumps, to operate closer to their most-efficient operating point and conserve energy. Staging can include the following techniques:

1. Constructing multiple small unit processes rather than single larger ones permits process requirements to match flow and load requirements by taking units out of service. For example, constructing several parallel bioreactors allows basins to be removed from service and reduce biomass, which reduces the oxygen requirements and associated aeration energy cost.
2. Staging of construction to match capacity requirements permits energy-consuming equipment to operate closer to its most efficient operating point.

Constructing multiple smaller parallel unit processes or equipment services allows for a better match between seasonal (e.g., tourist) variation in loads and infrastructure requirements. Beyond seasonal load variations, permanent load variations caused by industry coming online may also be managed through staged expansion of treatment capacity.

The benefits of staging can be difficult to quantify, because considerable monitoring, benchmarking, and estimating of costs associated with unimplemented alternatives must be completed to allow a comparison. In addition, it can be more difficult to estimate accurate construction costs for one large project than to determine the cost of mobilizing several smaller projects. Still, staging should be considered in many cases, and the complex economics should be evaluated as part of an energy evaluation decision process.

2.4.3.3 Covered Basin Heat Retention

Temperature can have a significant influence on biological wastewater treatment processes. This influences bioreactor size requirements and the mixed liquor suspended solids mass maintained in the system. The mass, in turn, influences the oxygen uptake rate, the oxygen requirements and, thus, the energy demand of the system.

Covering treatment basins can also reduce the loss of heat energy from biological reactors, allowing a higher temperature to be maintained in the biological process. Maintaining a higher temperature reduces basin size, increasing system capacity and saving cost.

In northern climates, equipment exposed to cold temperatures can freeze, requiring energy to thaw and maintain. Covering basins can avoid these energy costs.

2.4.4 Electromechanical Systems

Electrical motors convert electrical energy to the mechanical energy required to perform useful work. They consume a high percentage of the energy used in WWTFs, perhaps as much as 90%. From a 10-year life-cycle cost perspective, electrical motor operating costs can exceed capital costs by more than 50 times. For example, a USD$2,000 motor can consume USD$100,000 in electrical costs. Engineers should therefore focus on life-cycle cost when evaluating the economic feasibility of alternate electrical and associated systems.

This section provides a brief description of the electrical systems used in WWTFs in North America. Several of the references and reviewed documents, as well as numerous courses, provide background on electrical systems. In particular, the U.S. Navy developed an excellent training course on Electricity and Electronics (Jones, 1998) the first few chapters of which apply well to power systems; the remaining chapters apply to instrumentation and control systems (http://www.tech-systems-labs.com/navy.htm).

2.4.4.1 Electrical Motors

Types of Electrical Motors

Electrical motors can be classified into many different categories, for example:

- Size: Typically measured in horsepower
- Power: Alternate Current (AC) vs. Direct Current (DC)
- Phase: Single Phase vs. Three Phase
- Voltage: 110v vs. 480v or several other voltages
- Starting Torque Requirements: High or Normal
- Starting Current Requirements: High or Low
- Design: Squirrel Cage, Wound Rotor, Synchronous presents typical motors applied in WWTFs

In WWTFs, small motors typically use 110v single-phase power (less than one horsepower), while larger motors typically use 480v three-phase power. Exceptions exist for some larger, single-phase applications, such as metering pumps and certain heating, ventilating, and air conditioning components. The output power of a motor determines the rated capacity.

Squirrel cage induction motors typically provide the most appropriate service in industrial applications such as wastewater because they are low in cost and available in all power ratings and synchronous speeds. Voltage in this motor is applied directly to the primary winding, or stator, while the secondary winding, or rotor, consists of conductive metal bars (aluminum or copper) connecting in a conducting ring arrangement resembling a squirrel cage. The rotor windings comprise a complete electrical circuit, eliminating the maintenance of electrical connections to the rotating side of the motor. Squirrel cage induction motors produce torque through slip, or a difference between the operating speed and the synchronous speed. The alternating current in the stator rotates at a speed proportional to the AC frequency and the number of poles. The rotor attempts to follow this synchronous speed but falls slightly behind. Known as slip, this produces the torque in the attempt to reach equilibrium. Variable frequency drives (VFD), which are discussed later in the document, apply this principle to control motor speed.

Wound rotor induction motors require slip rings and brushes to apply voltage to the secondary windings, which require more maintenance. Speed reduction occurs, however, without the requirement for a VFD. This limits the application of wound rotor motors to special applications, such as hoists and cranes, which require high starting torques and intermittent operation.

Because synchronous motors apply direct current to the rotor, slip does not occur, and the rotor follows the magnetic field developed in the stator. This limits the torque that can be developed, requiring special starting mechanisms. Compared to asynchronous motors, for which the power factor varies based on load or size, synchronous motors provide the benefit of a controllable power factor. The power factor is the ratio of the actual power to the apparent power caused by inductive loads in an electrical system. As described later, electrical utilities surcharge for low power factors. Large synchronous motors in a power system allow for the adjustment of the overall power factor by changing the direct current to synchronous motors and the power factor to leading, lagging, or unity. Large synchronous motors can therefore maintain the overall system power factor to close to unity, avoiding any electrical utility surcharges.

Motor/Load Matching

Motors must be able to drive the intended load, often throughout a variable load range. Many motor designs and configurations can accomplish this goal; however, few of them can accomplish it efficiently over the entire load range. Any device that transforms one type of energy into another produces losses in energy such as heat, including electrical motors. For an electrical motor, efficiency is the ratio of the mechanical power output (Watts) to the electrical power input.

Standard induction motor efficiency increases with motor size and remains constant with variable load, generally varying by less than 5% over the possible load range. Power factor, however, does not remain constant with a reduction in load and can decrease by as much as 25% from full load to half load. As discussed above, electrical utilities can assess surcharges for low power factors, making it beneficial to match the motor to the load across the entire operating range.

Motor and load matching can require complex economic evaluations in an evolving inductive electrical system, as loads change during the course of different timescales. In some cases, it may be beneficial to replace motors if they are having a considerable effect on the overall system power factors. In other cases, the replacement cost may not be justified. A complex electrical system requires monitoring as changes occur and attention is required to determine the most cost-effective configuration to produce the required work.

Motor Efficiency Standards

Previous sections discussed how motor operating costs outweigh capital costs by many times and how motor efficiency does not vary considerably over the operating range. The maximum delivered motor efficiency varies based on the motor design: with higher-efficiency motors having higher-quality materials, precise manufacturing tolerances, higher-quality bearings, and long, metallic windings. Various organizations promote motor energy efficiency, including:

- Department of Energy – Energy Efficiency and Renewable Energy Program
- U.S. Environmental Protection Agency
- Independent Energy Companies
- National Electrical Manufacturers Association
- Institute of Electrical and Electronic Engineers (IEEE)
- Consortium for Energy Efficiency

WEF's Energy Conservation in Water and Wastewater Treatment Facilities (WEF, 2009) describes these organizations in detail and the programs offered to improve motor efficiency.

2.4.4.2 Variable Frequency Drives

The maximum output speed, torque, or power performed by a driver, typically an electrical motor, frequently does not match that required by the driven equipment. Although various techniques exist to transform the driver output power into the required power, the VFD is the most widely applied and offers the greatest potential in energy savings. Early adjustable speed drive devices, used to control driver output such as throttling valves and liquid rheostats, sacrificed efficiency to satisfy drive speed or power requirements. The development in the 1980s of the insulated-gate bipolar transistor, a critical modern VFD component, permitted low-cost, highly-efficient matching of motor/equipment speed and power requirements, and quickly became standard practice in industry and in WWTFs. Unlike wasteful techniques that do not perform useful work, VFDs match utility energy consumption to system energy requirements.

VFDs operate by reducing the utility power frequency [60 hertz (Hz) in U.S., 50 Hz in most other parts of the world]. The speed of an asynchronous induction motor varies proportionally to this frequency. The three main circuits in a VFD include:

1. Rectifier – Converts the input utility standard frequency AC voltage into DC voltage
2. Inverter – Converts the rectified DC voltage back to variable frequency AC
3. Regulator – Controls the rectifier and regulator to produce the desired frequency and voltage

The simple example presented in Table 2-6 compares a 20-hp motor running 24 hours a day at constant speed to a VFD-controlled motor, in a best-case scenario ignoring VFD inefficiency with power at a cost of USD$0.11/kWh.

This simple example demonstrates that the use of the VFD instead of a constant speed drive achieves a savings of USD$3,994. Actual economic analysis should consider the use of multiple smaller size units, the efficiency of the VFD across the service speed and load range, as well as the performance curve of the driven equipment (for example, the pump and system curve). Priced at between USD$50 and USD$200/hp (depending on the application), VFDs typically pay back their value within a couple of years of purchase, and should be the standard of comparison, rather than outdated, constant-speed applications.

Table 2-6. Simple VFD Savings Example.

Duration hrs/day	Speed % of full	Constant Speed Energy hp hrs	Constant Speed Cost USD$/day	VFD Energy hp hrs	VFD Cost USD$/day
2	100%	40	3.28	40	3.28
3	90%	60	4.92	54	4.43
5	80%	100	8.21	80	6.56
7	70%	140	11.49	98	8.04
4	60%	80	6.56	48	3.94
3	50%	60	4.92	30	2.46
24		480	39.39	350	28.72

Energy savings: 130 hp hours/day; 96.98 kWh/day; 27%.

Cost Savings: 10.67 USD$/day; 3,894 USD$/year

2.4.4.3 Supervisory Control and Data Acquisition Systems

SCADA systems provide a central location for controlling, monitoring, and recording the performance of energy-consuming devices throughout an entire WWTF. They can be applied in a wide variety of configurations, for example:

1. Housing the logic for all control loops throughout a WWTF
2. Monitoring and changing set points for remote control loops housed in individual controllers, programmable logic controllers
3. Monitoring status of equipment operated remotely
4. Operating remote manual equipment
5. Archiving equipment operation status, measurements of primary elements (flow meter measurements, for examples), and power draw of individually metered equipment

These represent just a few examples of the wide variety of SCADA configurations, depending on owner preferences and the evolution of a system within a particular WWTF. In many cases, more than one SCADA solution applies to similar situations.

As discussed in other sections of this document, ECM opportunities exist through the application of control loops to equipment – for example, DO blower control. These loops may or may not be housed in a centralized SCADA system.

Energy management integrates into SCADA systems very effectively by integrating real-time energy data from the plant's electrical service or any sub-metering with billing schedules and operating set points. This allows plant operators to visualize the cause and effect of equipment operation in terms of energy cost to modify operating strategies and may allow the SCADA system to control certain operations, such as peak shaving to reduce energy costs.

SCADA systems facilitate operating strategies such as setting a target demand based on an engineering evaluation of the system and then monitoring the system to prevent this from being exceeded. If the target is exceeded and the demand charge increases, SCADA allows for an evaluation of the cause and justification for the increase. In its Quality Energy Efficiency Retrofits for Wastewater Systems (EPRI, 1998), EPRI describes the following SCADA features that could be integrated effectively with ECM:

- Real-time monitoring, load reduction, or demand load-shedding to take better advantage of time-of-use rate structures.
- Monitoring of real-time electrical energy prices from a variety of suppliers and selection of the lowest-cost power available as deregulation of energy markets proceeds
- Alarming to monitor high demand charges or equipment inefficiencies
- Equalization basin management for off-peak treatment to reduce aeration costs (an example of an ECM that reduces cost without reducing energy consumption)
- Managing equipment for efficient operation – for example, in a battery of parallel pumps all controlled with VFDs
- Return-activated sludge rate adjustment based on influent flow rate
- Blower DO control

A team responsible for energy management within a WWTF should evaluate SCADA energy management goals and operations periodically.

2.4.5 Pumping Systems

Pumping plays a crucial role in WWTFs and can consume a considerable portion of the energy, depending on the topography, the WWTF layout, and the need for intermittent or effluent pumping. Figure 2-1 presents energy consumption for a typical WWTF and indicates that pumping consumes 14% of the energy. In fact, this value varies considerably between facilities. Although WWTFs apply various types of pumps, this document focuses on centrifugal pumps, which are the most commonly applied and, therefore, offer the greatest opportunity for energy savings.

Similar to motor efficiency, pump efficiency is the ratio of the water power derived from the pump to the motor power input to the pump shaft. Pump efficiency varies considerably more than motor efficiency over the potential operating range of a pump.

Figure 2-5 presents typical pump characteristic curves for a single impeller diameter. The curve plots the flow rate against the total dynamic head. This plot includes a system curve, which characterizes the total dynamic head developed in the system in which the pump operates. Centrifugal pumps operate at the intersection of the pump curve and the system curve. Manufacturers design pumps with different operating conditions for different applications, but in this case, the efficiency increases to a maximum and then decreases. This pump suits this application well because the point of highest efficiency coincides with the pump operating point, not a difficult achievement in this static, illustrative example. Designing to maintain high efficiency with dynamic operation in a VFD-driven application can be more challenging. Operators should verify the performance of operating pump systems periodically to ensure they perform at the intended efficiency as a system ages and evolves.

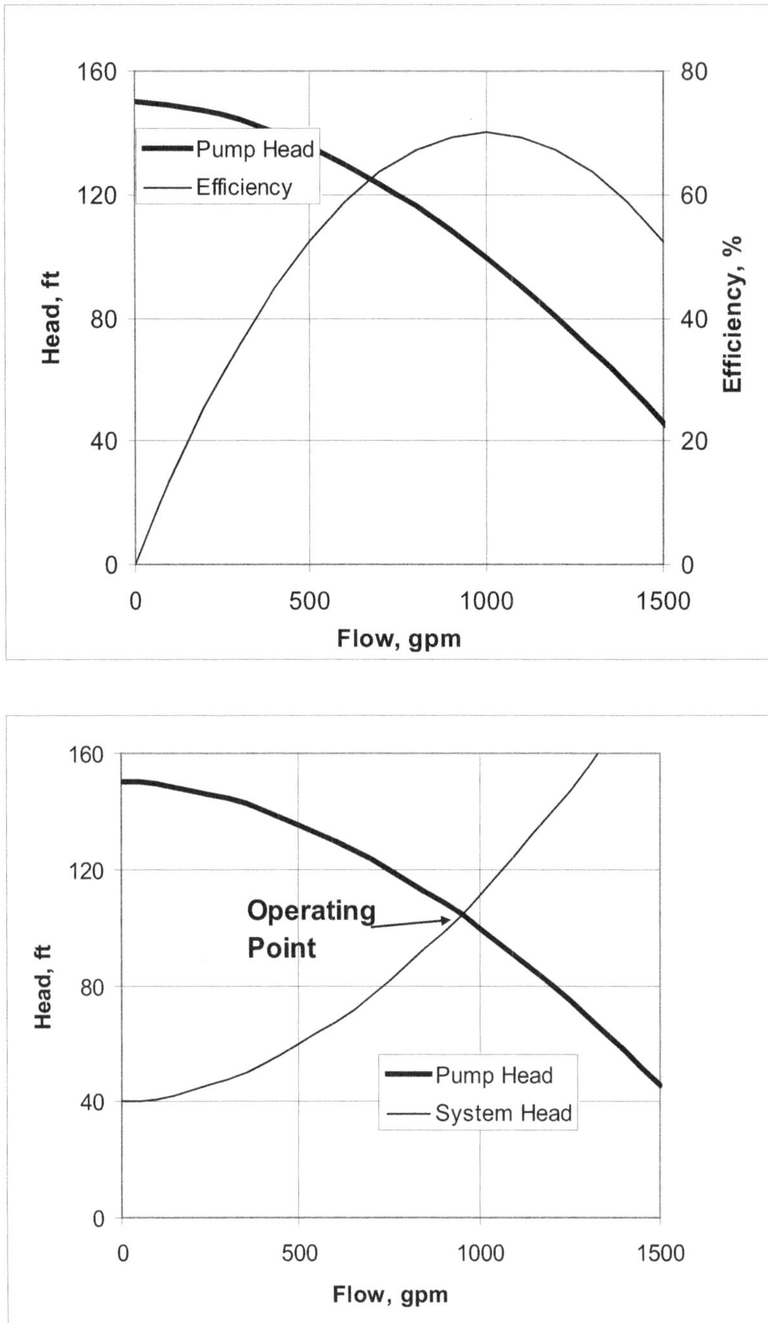

Figure 2-5. Typical Pump Characteristic Curves.

Reprinted with permission from the Water Environment Federation, MOP 32, 2009.

Large horsepower, multi-pump systems may warrant individual metering to monitor power consumption with pump efficiency and system monitoring and trending available in the SCADA system. Flow, suction and discharge pressure, run time, temperature, and more advanced maintenance monitoring (such as vibration analysis) might also be included for these expensive systems.

When pump efficiency does not meet expectations, physical and economic analysis often reveals the cause of deviations and determines the cost-effectiveness of corrective action. Corrective actions may include:

- Pump testing to verify and collect accurate data
- Shaft alignment
- Rebuilding of pump, including:
  - Replacing the impeller – impellers age and can be damaged by cavitation
  - Replacing the pump with a more suitable selection
- Optimization of the pumping system – for example, by modifying the control system or VFD operation

In cases where each percentage increase in pump efficiency conserves considerable energy and/or power cost, pump coating systems may prove cost-effective. The hydrophobic, smooth-surfaced polymer coats can provide up to a 5% increase in efficiency. Evaluating pump coating performance requires expertise to determine effectiveness for specific application.

2.4.6 Aeration System

Figure 2-1 indicates that wastewater aeration consumes greater than 50% of WWTF energy, far more than any other category. WWTF energy optimization should therefore begin with an evaluation of the efficiency of the aeration system, as it offers the greatest opportunity to minimize energy consumption, GHG emissions, and cost. Table 2-7 provides typical aeration energy usage for different sized WWTFs. The table assumes 133 kWh/ML/d for aeration at all flows.

Table 2-7. Aeration Energy Usage.

Capacity	Electricity kWh/day
4 ML/d (1-mgd)	532
20 ML/d (5-mgd)	2,660
40 ML/d (10-mgd)	5,320
80 ML/d (20-mgd)	10,640
190 ML/d (50 mgd)	26,600
380 ML/d (100-mgd)	53,200

The microorganisms employed in biological treatment systems require oxygen for the following:

1. Aerobic degradation of biochemical oxygen demand (BOD)
2. Nitrification of ammonia

Microorganisms obtain oxygen from the DO in solution, which is accomplished through several different types of systems. The cost of aerating wastewater is so high because oxygen does not dissolve readily in water. The saturation concentration of oxygen in water also depends on temperature, with the saturation concentration decreasing with temperature.

2.4.6.1 Fine-Pore Diffusers

Table 2-8 (WEF, 2009), presents aeration efficiency for various aeration devices. Over the past 25 years, the application of alternate aeration systems has determined that fine-pore aeration systems typically provide the most-efficient aeration. Most large WWTFs have migrated toward this type, although some justified exceptions do exist.

Table 2-8. Aeration Efficiency.

Aerator Type	Aeration Efficiency, kg O_2/kWh (lb O_2/hp hr)	
	Standard Conditions	Field Conditions
Fine-pore aeration[1]	5.0 – 6.5 (8.2 – 10.7)	2.5 – 3.5 (4.1 – 5.8)
Course bubble aeration[1]	2.5 – 3.5	1.0 – 2.0
Surface centrifugal (low speed)	1.2 – 3.0	0.7 – 1.4
Surface centrifugal (draft tube)	1.2 - 2.8	0.7 – 1.3
Surface axial (high speed)	1.2 – 2.2	0.7 – 1.2
Downdraft open turbine	1.2 – 2.4	0.6 – 1.2
Downdraft closed turbine	1.2 – 2.4	0.7 – 1.3
Submerged turbine, sparger[1]	1.2 – 2.0	0.7 – 1.1
Submerged impeller	1.2 – 2.4	0.7 – 1.1
Surface Brush and blade	0.0 – 2.2	0.5 – 1.1

Source: Based on WEF MOP FD-13 Table 5.1 with additional content
Note: 1.0 kg/wWh = 1.66 lb/(hp hr)
1 – Includes blower power requirements.

The economic benefits of fine-pore aeration vary, depending on the application; however, energy savings between 20-75% are typical. Again, a life-cycle cost analysis must be completed to determine the cost-effectiveness of converting a system to fine-pore aeration. For typical new systems and systems which require replacement at the end of useful life, fine-pore aeration usually provides the most cost-effective life-cycle solution.

The decision to select ceramic or membrane fine-pore diffusers is also site-specific. Ceramic diffusers tend to biologically foul more than membrane diffusers.

Fine-pore aeration systems may not be the most cost-effective alternative in the following cases:

- Short solids retention times (SRT), carbonaceous, or high-rate activated sludge systems in which low aeration transfer efficiency occurs because of the presence of surfactants (which line the air-water interface, preventing air transfer). In these cases, the benefits provided by fine-pore aeration may be minimal and the additional capital cost not justified.
- Shallow aeration basin applications
- Industrial applications or applications in which significant diffuser fouling can occur

2.4.6.2 Blowers

Blowers are usually more efficient than compressors at providing air to diffused aeration systems, because the relatively high air flow rate and low discharge pressure better suits a blower. WWTFs typically employ three types of blower systems:

1. Positive displacement (PD) blowers
2. Multi-stage centrifugal blowers
3. Single-stage centrifugal blowers

In the basic sense, blowers operate in a similar manner to pumps by imparting kinetic energy to a fluid to provide pressure and velocity. While pumps act on an incompressible fluid (water) blowers act on compressible fluid (air); the airflow pressure and temperature conditions affect the volume. Volumetric actual flow rate is expressed as cubic meters per hour (m^3/h) in most parts of the world and in cubic feet per minute (ft^3/min) in the U.S. Standard conditions of 1 atmosphere, 20ºC (68ºF), and 36% relative humidity represent units of normal m^3/h or standard ft^3/min. Because of the defined properties and density, the units actually represent a mass flow rate at standard conditions.

Positive Displacement Blowers

Some smaller WWTFs employ rotary lobe PD blowers. These blowers discharge an identical volume of air, the blower displacement with each rotation of the blower shaft, regardless of discharge pressure. Blower efficiencies of less than 60%; however, frequently make PD blowers uncompetitive compared to centrifugal blower alternatives.

Multi-Stage Centrifugal Blowers

These blowers function much like a system of pumps operated in series, with each successive stage increasing the pressure. Multi-stage blowers have been used in WWTFs for many years and provide long, useful, service life. Multi-stage blower efficiencies typically range from 60-75%, with limited turndown capabilities of 60-70% of capacity compared to single-stage blowers. The limited turndown requires configurations with several blowers operated in series, as well as control systems to start and stop blowers automatically as system air requirements change.

Single-Stage Centrifugal Blowers

Single-stage blowers employ only one impeller to produce airflow operating at higher speeds than a multi-stage blower. Inlet and discharge guide vanes pre-rotate the air before it enters the impeller, which allows single-stage blower efficiency to range from 65-85% and can remain close to best efficiency across a wide range. Turndowns to 40% of full capacity can also be achieved.

High-speed, single-stage centrifugal turbo blower that employs a magnetic or air bearing (which require very little maintenance) in a configuration similar to an aircraft jet engine. Turbo blowers achieve efficiencies and turndowns similar to other single-stage configurations while requiring less space and producing less noise.

2.4.6.3 DO and Blower Control Systems

The oxygen requirements of an activated sludge system change depending on many factors, such as the influent BOD, the influent ammonia, the system SRT, temperature, and others. The system therefore responds to a constant airflow by varying DO levels, increasing and decreasing over the course of the diurnal, seasonal, and other oxygen demand variations. To maintain adequate residual DO during peak oxygen requirements, a high airflow rate would be required. However, this high aeration rate would also result in high DO levels in periods of low oxygen demand, wasting energy as a result.

Manual aeration control offers one alternative for controlling DO levels in activated sludge systems. To this end, operators can adjust the blower speed with a VFD in response to the basin DO measurement, thus changing the airflow to the basin. Although an improvement over constant airflow delivery, this alternative requires constant attention for adjustment, as the oxygen requirements change, increasing labor cost. Automatic DO control systems deliver the correct amount of oxygen without excessive residual by monitoring DO in the basin with a DO probe, and controlling the airflow to the basin with airflow control valve. A signal from a DO probe strategically placed in an aeration basin may control either a single valve or several control valves in parallel aeration basins of the same configuration, assuming that an accurate flow split occurs and that the basins have similar oxygen uptake rate profiles.

Because the blower seeks for the appropriate speed as aeration requirements change, blower control does not work effectively by incorporating DO measurement to control blower speed directly. Blower control operates indirectly and is based on achieving a desired pressure in the blower discharge piping. If the aeration system calls for more air, the pressure drops and blower speed increases. If the aeration system calls for less air, the pressure increases and blower speed decreases.

Compared to manually controlled aeration systems, automatic DO control systems conserve 20-40% of aeration energy. They usually prove cost-effective for new activated sludge installations. Again, a life-cycle cost assessment should be conducted when modifying existing systems.

2.4.7 Energy Recovery Systems

WERF developed a summary of technologies applied for energy recovery from WWTF solids (WERF, 2008), including:

- Sludge-to-biogas
- Sludge-to-syngas
- Sludge-to-oil
- Sludge-to-liquid

Mesophilic anaerobic digestion of primary and waste-activated sludge (WAS) producing methane gas recovers energy in the majority of energy recovery systems operated in the United States. This section discusses methane gas production through anaerobic digestion and the systems that convert the gas to energy. Several types of energy recovery systems are discussed in Chapter 3.0 of this report.

Typical energy recovery systems typically provide approximately 20–40% of the energy requirements for an activated sludge WWTF, depending on the type of treatment technology employed and the treatment level. Table 2-9 presents digester gas utilization technologies and the equipment they require.

Table 2-9. Digester Gas Utilization Technologies.

Use	Equipment
Digester Heating	Boiler, Heat Exchangers, Heat Recovery Equipment
Electrical Power Generation	Gas Cleaning, IC Engine, Microturbine, Fuel Cell, Sterling Engine, Steam Turbine
Building Heating	Heat Recovery Equipment, Heat Exchangers
Air Conditioning	Heat Recovery Equipment, Chiller
Biosolids Drying	Dryer, Heat Recovery Equipment
Biosolids Pasteurization	Boiler, Heat Recovery Equipment, Heat Exchangers
Thermal Hydrolysis	Boiler, Heat Recovery Equipment, Direct Heat Injection
Methane Gas Retail	Gas Treatment
Methane Gas Vehicles	Gas Treatment
Pump/Blower Drive	Gas Engine

Reprinted with permission from the Water Environment Federation, MOP 32, 2009.

Anaerobic digestion produces approximately 0.075 m^3 of digester gas per m^3 of wastewater treated (10,000 ft^3 of digester gas per million gallons). Digester gas contains between 40% and 75% methane (CH_4), with 60% being common, resulting in a higher heating value of 22,000 kJ/m^3 (600 Btu/ft^3).

Table 2-10 presents electrical and thermal efficiencies for energy recovery systems employed to harness the gas. These systems typically operate with an electrical efficiency of 30%, resulting in electrical production of 15 kWh/m^3/s (350 kWh/MG). Energy recovery systems become cost-effective at WWTF capacities of approximately 1.0 m^3/s (approximately 20 mgd), although a site-specific analysis must be conducted. New technologies being developed and discussed later in this document may become effective at smaller WWTF capacities.

Table 2-10. Gas Utilization Equipment Efficiency.

Technology	Net Electrical Efficiency		Net Thermal Efficiency		Size Range kW
	Range %	Typical %	Range %	Typical %	
Internal Combustion Engine	25 – 45	33	40 – 49	40	50 – 5K
Internal Combustion Engine – Lean Burn		37			
Gas Turbines	23 – 36	30	40 – 57	40	250 – 250K
Microturbines	24 – 30	27	30 – 40	35	30 – 250
Steam Turbine	20 – 30	25	20 – 45	45	500 – 1,300K
Stirling Engine	25 – 30	27	45 – 65	60	1 – 50

Reprinted with permission from the Water Environment Federation, MOP 32, 2009.

Although Table 2-10 provides both electrical and thermal efficiencies, achieving both simultaneously in a CHP requires a detailed engineering analysis. Electrical and thermal efficiency change inversely proportional to one another, so that increasing the electrical efficiency decreases the possible thermal efficiency. Electrical energy can be used internally in the plant or be exported to the electrical supply grid. Thermal energy could be used to heat digesters, dry sludge, or heat buildings.

CHAPTER 3.0

CASE STUDIES OF NOVEL ENERGY OPTIMIZATION/ENERGY RECOVERY TECHNOLOGIES AND/OR PRACTICES

3.1 Introduction

The following sections document in a series of case studies novel (yet full-scale proven with at least one operating installation) energy optimization/energy recovery technologies or practices. Case study documentation follows the guidelines defined by Black & Veatch, the GWRC umbrella contractor, titled *UkWIR Energy Efficiency Research Project CL 11, A Compendium of Best Practice and Case Studies, Final Document – Case Studies and Examples, Selection Criteria and Data Collection Guidance, Revision 3*.

The documentation guidelines developed for the GWRC project provides information on a wide range of energy optimization/energy recovery options in new and existing facilities. As such, the information collected is based on verifiable, full-scale sources, so that results can be quantified and any implementation risks or obstacles can be identified and managed. In general terms, the information provided in these case studies includes a general overview of the technology/practice (independent of the specific case study), and for a particular facility, it establishes the background, the type and size of the process, the situation before and after any changes, the changes themselves and the results obtained.

3.2 Advanced Anaerobic Digestion and Combined Heat and Power – Columbus Biosolids Flow Through Thermophilic Treatment Technology (CBFT3)

- The project consists of the following main elements:
  - An advanced anaerobic digestion system known as the CBFT3 process.
  - A CHP system for converting anaerobic digester gas into renewable green power.
  - A fats, oils, and grease (FOG) receiving and processing system which improves digestion and anaerobic digester gas production.
- The purpose of the CBFT3 process is to convert sewage sludge to Class A biosolids and to maximize anaerobic digester gas production for conversion into renewable energy.
- The CBFT3 process consists of a continuous-feed, stirred-tank reactor followed by a 30-minute batch detention step with both tanks operating at thermophilic temperatures.
- The 30-minute batch occurs within a newly developed and innovative plug-flow reactor.

- The plug flow reactor is operated in conjunction with mesophilic and thermophilic anaerobic digesters.
- The CHP system consists primarily of two 1.75 MW internal combustion engines used for generating electric power and thermal energy.
- The FOG receiving and processing system consists of a 340 m^3 (12,000-gallon) tank used to receive, mix and heat hauled grease trap waste prior to feeding the treated material to the thermophilic digester.

3.2.1 Potential Benefits

- Considerably shortens digestion batch times, thus reducing the amount of digester volume required.
- The CHP system converts anaerobic digester gas into thermal and electrical energy.
- The facility is the first thermophilic digestion process in the United States run entirely off thermal energy captured by a CHP system.
- The FOG receiving and processing system improves overall digester gas production by about 50% while increasing sludge reduction by about 10%.
- It is estimated that the CBFT3, CHP, and FOG systems will result in a net carbon footprint reduction of 9,600 metric tons per year of CO_2 equivalent, primarily as a result of purchased electricity offsets.

3.2.2 Concerns

- The CBFT3 process is the first of its kind with limited operational experience to this point.
- The CHP system's internal combustion engines are sensitive to digester gas contaminants, in particular hydrogen sulfide and siloxanes. A fuel treatment system has been included to protect the engines from potential damage and efficiency loss.
- The CHP engines require continual cooling. This has also been addressed and included in the project.

3.2.3 Range of Potential Savings

- CBFT3 enhancements were constructed for about USD$3 million less than a comparable U.S.EPA "time-and-temperature" batch system because of the reduction in required batch detention time from 24 hours to only 30 minutes at 55ºC.
- Approximately 40% of the facility's electrical demand is offset by power generated by the CHP system.
- Based on electric energy savings at USD$0.075/kWh, payback is estimated at less than 10 years.

3.2.4 Application Potential

- The CBFT3 process is the first of its kind. A detailed record of its performance will help in moving this technology forward and into other WWTFs.
- Internal combustion engines are a time-tested and proven technology and are the centerpiece of hundreds of successfully operating CHP systems around the world.
- The addition of FOG to anaerobic digestion systems has become very popular in recent years with many more planned for the immediate future.

The figure below depicts the CBFT3 process and CHP system with the addition of FOG to the thermophilic digester.

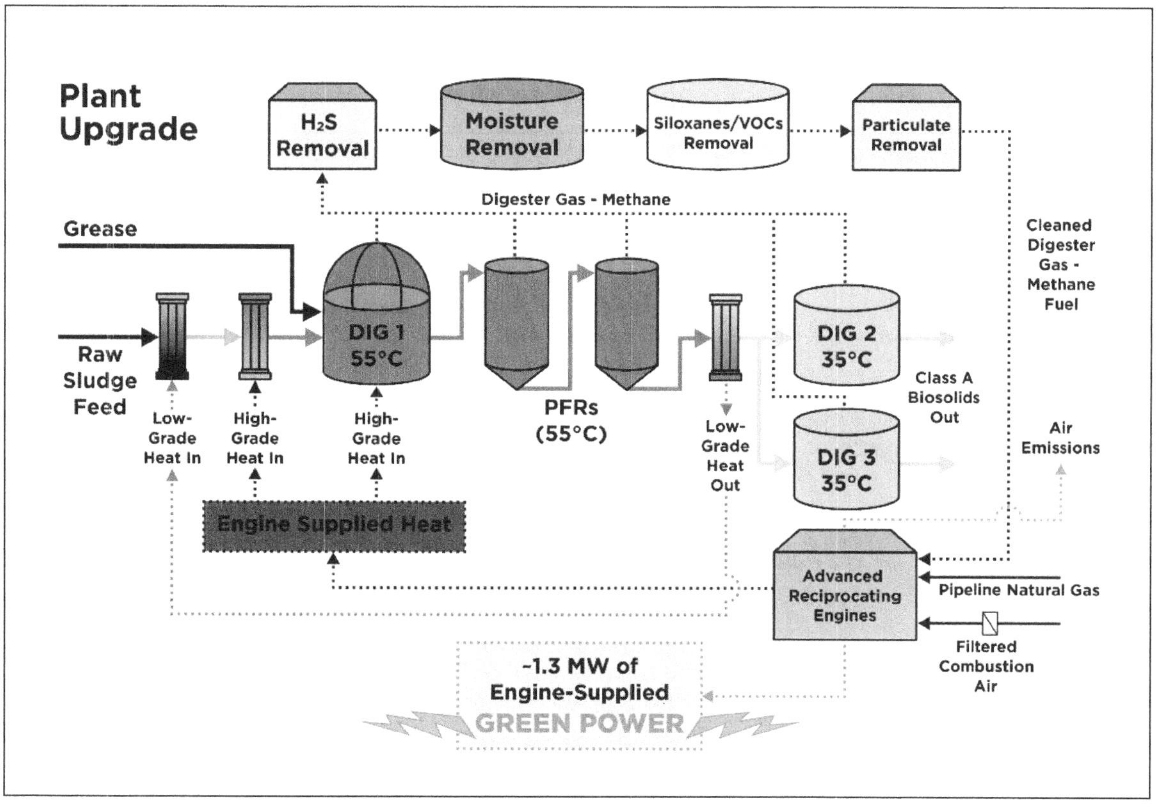

Figure 3-1. CBFT3 Process and CHP System with the Addition of FOG to the Thermophilic Digester.
Reprinted with permission from Brown & Caldwell, 2010.

3.2.5 South Columbus Water Resource Facility, Columbus, GA, U.S.

Table 3-1. Advanced Anaerobic Digestion and CHP – CBFT3 Technology.

Ref	Enquiry Item	Response Information, Description and Remarks
1	Location: country, urban or rural:	Columbus, Georgia, a city located on the Chattahoochee River on the western border with Alabama. Columbus is the third largest city in Georgia.
2	Sector: clean, waste or sludge:	Sludge.
3	Works owner or operator: with financial set-up, regulatory or not.	Columbus Water Works: Wastewater conveyance and treatment service provider, not regulatory.
4	Size: flows and loads or population equivalent:	Columbus Water Works provides wastewater collection and treatment to 62,000 locations serving a population of approximately 200,000. The project is at the South Columbus Water Resource Facility (SCWRF) which currently treats 132,500 m^3/d of sewage.
5	Energy provider: with costs, incentives, taxes and conditions:	Georgia Power/Southern Company. Average rate is USD$0.073 per kWh.
6	Process: physical, chemical, or biological description:	Biological; secondary treatment with anaerobic sludge digestion.
7	Component: all or part of the works.	Part of the works.

Table 3-1. Advanced Anaerobic Digestion and CHP – CBFT3 Technology, continued.

Ref	Enquiry Item	Response Information, Description and Remarks
9	Process/Plant changes: mechanical, electrical or controls:	Power generation from biogas is accomplished by two state-of-the-art, advanced reciprocating internal combustion engines; each rated at 1.75 MW. New electrical switchgear will synchronize the renewable energy onto the plant grid. Digester gas storage and treatment upgrades are also being provided.
10	Civil/physical changes: to water/effluent quality, civil works, or process:	Average volatile solids reduction by mesophilic digestion is expected to be about 58%. Average volatile solids destruction by thermophilic digestion is expected to be about 65 to 68%.
11	Operational changes: skill levels, procedures and maintenance routines:	The digestion process and gas treatment will not require additional skills beyond that exhibited by current plant staff. The engine-generators will be maintained by the engine manufacturer.
12	Risks and dependencies: risk assessment of project and changes.	The CHP system will require more attentive operation and maintenance of gas treatment system to ensure high-quality gas feed to engines. Digester-gas-fueled internal combustion engines are a proven process. Digestion improvements will provide greater operational flexibility.
13	Implementation: design, build, procurement, installation and commissioning:	Construction of full-scale CBFT3 improvements and CHP system began December 2007, with start-up expected in mid 2010.
14	Energy efficiency gains: kWh & kWh/m^3	Average gas production is about 9,200 m^3/d. Engine electrical efficiency is about 38%. Thermal efficiency is about 42%. Combined overall efficiency is about 80%. Co-generation expected to produce 1.2 to 1.4 MW electrical power, on average. Engines can produce 40 to 50% WWTF power requirements.
15	Cost/benefit analysis: financial appraisal or payback time.	Base payback is estimated at 9.6 years (no carbon credits and electric energy savings at USD$0.075/kWH)
16	Project review: could it be improved or developed?	Project included several first-of-its-kind engineering applications. Full-scale operational experience will lead to opportunities for further refinement.
17	Confidence grade: on data provided.	High.

3.2.5.1 Observations

Figure 3-2. SCWRF Plug Flow Reactors during Construction.
Reprinted with permission from Brown & Caldwell, 2010.

3.3 Co-digestion of Dairy Manure with WWTF Sludge

3.3.1 Technology Overview

3.3.1.1 Process Description

- Methane is a particularly potent GHG with a global warming potential 23 times higher than CO_2. Capturing CH_4 through anaerobic digestion of manure allows for its use as an alternative to natural gas in combustion and power production.
- Anaerobic digestion is a naturally occurring biological process in which a consortium of anaerobic bacteria converts organic material into CH_4 and CO_2 in the absence of air. More than 70% of WWTFs across the U.S. use anaerobic digestion for stabilizing organic matter in wastewater solids, reducing pathogens and odors, recovering nutrient, and reducing the total solids/sludge quantity while producing biogas. Anaerobic digestion is also a commonly-used technology for stabilization of animal wastes and manure.

Figure 3-3. SCWRF Engine-Generators during Installation.
Reprinted with permission from Brown & Caldwell, 2010.

- Anaerobic digestion is also a commonly-used technology for stabilization of animal wastes and manure.
- Co-digestion is a relatively new concept in which two or more substrates are simultaneously digested in anaerobic digesters. A main substrate (e.g. manure or sewage sludge) is typically mixed and digested together with minor amounts of a single substrate or a variety of additional substrates.
- Three major drivers promote co-digestion:
  - Digesters in WWTFs are usually oversized. Addition of co-substrates helps to produce more gas and, consequently, more electricity at only marginal additional cost. The extra electricity produced may cover the energy needs of wastewater treatment at a reasonable cost.
  - Co-digestion of manure and certain type of organic wastes (i.e., cheese whey, vinegar waste) may provide better conditions for acidification and dilution, which can improve digestibility.
  - Agricultural biogas production from manure alone (which has a relatively low gas yield) is economically not viable at current oil prices. Addition of co-substrates with a high CH_4 potential not only increases gas yields but potentially increases the income through tipping fees.
- Despite the potential benefits, co-digestion may require additional facilities for pretreatment, gas-handling, and treatment; increase solids loading to dewatering facilities; and may have impacts on recycle stream, mixing, and heating requirements as well as the quality of biosolids produced.
- In 2002, about 2,000 agricultural plants were in operation in Germany, most of them co-digestion facilities. Considerably fewer were in operation in Austria (110),

Switzerland (71), Italy (> 100), Denmark (>30), Portugal (>25), Sweden, France, Spain, England, and some other countries.

- The exact number of U.S. facilities that are co-digesting manure with other substrates is unknown. California currently has about 22 biogas-producing digesters located on dairy farms. A process schematic of a typical manure co-digestion facility is presented in Figure 3-4.

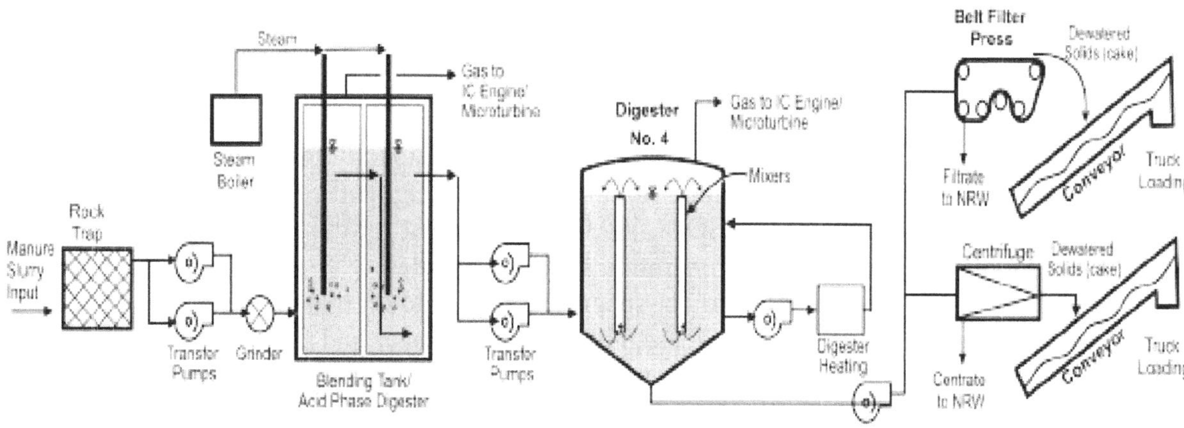

Figure 3-4. Process Schematic of a Typical Manure Co-digestion Facility.
Reprinted with permission from CH2M HILL.

3.3.1.2 Potential Benefits

Manure co-digestion offers the following potential benefits over conventional anaerobic digestion[1], unmanaged manure[2], and non-renewable energy[3]. The benefits are as follows

- Improvement on digestibility and biogas production[1]
- Source of revenue from tipping fees[1,2,3]
- Production of biogas, a renewable (inexhaustible) energy that reduces the reliance on non-renewable (exhaustible) fossil fuel power[3]
- Reduction in GHG emission due to capture of CH_4 and nitrous oxide (NOx) that would have been emitted by the manure and are collected instead[2,3]
- Relatively low carbon footprint technology (50 g/kWh vs. 950 g/kWh coal) [3]
- Co-digestion projects require modifications and some plant upgrades in the existing infrastructure, which is relatively quick and inexpensive compared to construction of a new coal or nuclear power plant[3]
- Reduction in ammonia emission through removal of manure from stockpiles and, therefore, prevention of ammonia emissions[2]
- Reduction in odor and improvement in groundwater quality because of waste management practices[2]

3.3.1.3 Concerns

Contrary to the benefits mentioned above, there are a few concerns associated with the implementation and operation of co-digestion that may require mitigation.

- Increased nitrogen oxide (NOx) and sulfur oxide (SOx) emissions caused by combustion of the extra biogas from the manure digestion

- Successful and reliable delivery of manure to the digester facility
- Requirement for additional facilities for pretreatment, gas handling, and treatment
- Increased solids loading to dewatering facilities; and may have impacts on mixing, and heating requirements of digesters as well as the quality of biosolids produced
- Inappropriate dilution and loading may result in digester failure
- Transporting and holding additional waste in the field may impose aesthetic problems (i.e., odor)
- Recycle streams (when returned to the head of the WWTF) may impact WWTF performance, potentially causing a plant upset and permit violations

3.3.1.4 Range of Potential Cost/Savings

Project cost is highly case-specific and dependent on capacity for co-digestion, upgrade needs for pretreatment and manure holding, gas cleaning, and co-generation facilities – for example, approximately USD$1,400,000 investment for the manure digester (3,200 m^3), gas cleaning, and power generation system upgrades.

Cost savings are also highly project-specific and depend on manure (waste) collection and transportation fees, tipping fees (if available), digester and gas system operational and maintenance (O&M) costs, electricity unit cost, renewable energy credits or "Green Tags" and GHG emission reduction credits. Future manure co-digestion projects may also receive particulate matter emission reduction credit. In addition, the *American Recovery and Reinvestment Act* (ARRA) of 2009 includes a 3-year extension of the renewable energy production tax credit (PTC) and a new program that allows renewable energy developers the option to forgo the PTC in lieu of secure a grant from the U.S. Treasury Department in the amount of a 30% investment tax credit.

3.3.1.5 Application Potential

The co-digestion concept has been successfully applied in Germany, Austria, Denmark, and other European countries for more than three decades. In 2006, there were about 75,000 dairy farms with more than nine million dairy cows in the U.S., with the majority of the dairy farms being located in Midwest. California is home to about 1,800 dairies that represent over 1.7 million dairy cows. The resulting manure is a significant source of CH_4 for use as a renewable energy source. Capturing CH_4 through anaerobic digestion of manure allows for its use as an alternative to natural gas in combustion and power production. The California Air Resources Board estimates that manure management projects utilizing anaerobic digestion could eliminate up to one-million metric tons of carbon dioxide equivalent (mT CO_2 e) by 2020 (Anders, 2007). There is potential for co-digestion implementation in locations where manure and food waste and federal- and state-supported incentives are available and anaerobic digesters have extra capacity to accommodate additional wastes are.

3.3.2 Co-digestion of Dairy Manure with WWTF Sludge Case Study: Inland Empire Utilities Agency (IEUA) Regional Plant 1 (RP-1) Co-Digestion of Dairy Manure to Energy, City of Ontario, California, U.S.

Table 3-2. Co-digestion of Dairy Manure with WWTF Sludge Case Study: IEUA RP-1 Co-digestion of Dairy Manure to Energy, City of Ontario, California, U.S.

Ref	Enquiry Item	Response Information, Description and Remarks
1	Location: country, urban or rural	Urban area in industrial setting
2	Sector: clean, waste or sludge	Wastewater
3	Works owner or operator: with financial set-up, regulatory or not	**Owner and Operator:** WWTF and co-digestion facilities are owned and operated by Inland Empire Utilities Agency. Organization set-up: regulated public agency Source of revenue: revenue is realized by charging user fees for wastewater collection and treatment services provided.
4	Size: flows and loads or population equivalent	Permitted Capacity is 167,000 m^3/d (60 mgd) Average Daily Flow is 227,000 m^3/d (44 mgd) Population Equivalent of Plant Capacity is 600,000 capita Average Daily Loads: BOD: 34,400 kg/d TSS: 35,000 kg/d TKN: 6,600 kg/d
5	Energy provider: with costs, incentives, taxes and conditions	The daily average electrical demand to operate CCWRF is around 6.0 to 7.0 megawatts (MW). The WWTF buys electricity from Southern California Edison and Coral Energy LCC (Certified Independent System Operator).The 2009 cost for electricity is between USD$0.12 per kWh.
6	Process: physical, chemical, or biological description	Liquid Treatment Process: primary treatment, secondary treatment, filtration and disinfection.

Table 3-2. Co-Digestion of Dairy Manure with WWTF Sludge Case Study: IEUA RP-1 Co-Digestion of Dairy Manure to Energy, City of Ontario, California, U.S., continued.

Ref	Enquiry Item	Response Information, Description and Remarks
7	Component: all or part of the works	Preliminary and Primary Treatment – screening, grit removal, chemically enhanced primary settling, primary effluent flow equalization. Secondary Treatment – activated sludge followed by secondary clarification. Tertiary Treatment – dual bed granular media filtration and chlorine disinfection. Solids Processing – primary sludge thickening via gravity thickening, WAS thickening via DAF. Thickened sludge are mixed and co-digested in three phase anaerobic digestion (mesophilic acid phase digestion followed by thermophilic gas phase digestion with a third stage unheated gas phase digestion. Dewatered sludge via belt filter presses is hauled to IEUA's co-composting plant in Chino, operated by Synagro. The biogas is cleaned and used in spark ignition lean-burn engines and microturbines to generate electricity.
8	Specific energy problem: including quality or consent details	Rising costs of electricity resulted in increased operational costs. The cost of electricity was the largest component of the WWTF operational costs. Agency's goal to become 100% self sufficient in producing energy from renewable energy sources to meet 11 MW of electrical energy needed for operating all IEUA facilities. Manure co-digestion is the one of the contributors to the agency's goal. In June 2001, the Commerce Energy Team awarded a program contract under the California Energy Commission's (Energy Commission's) Public Interest Energy Research (PIER) Renewable Program to conduct research on strategies for making renewable energy more affordable in California. As part of PIER Renewable Program, co-digestion of manure with food waste was pilot tested at RP-1.
9	Process/plant changes: mechanical, electrical or controls	No changes.
10	Civil/physical changes: water/effluent quality, civil works, or process	Construction of the pilot plant's materials handling system including installation of pumps, storage tanks, valves, metering, electrical and instrumentation and control systems, sampling ports, and related equipment. Construction of the pilot plant's digesters gas cleaning, and gas collection, piping, and safety systems upgrade, which included the following: Installation of new physical-chemical processes for hydrogen sulphite (H_2S) treatment involving new sponge media for H_2S absorption under anaerobic conditions and oxygen injection for operating the system under aerobic conditions. Expansion of gas collection system to accommodate increased gas production. Safety system upgrades to the existing piping systems. Piping modifications to connect the modified digesters with the existing gas distribution system.

Table 3-2. Co-Digestion of Dairy Manure with WWTF Sludge Case Study: IEUA RP-1 Co-Digestion of Dairy Manure to Energy, City of Ontario, California, U.S., continued.

Ref	Enquiry Item	Response Information, Description and Remarks
11	Operational changes: skill levels, procedures and maintenance routines	Food wastes including cheese whey, salad dressing and ice cream were brought to RP-1 and stored into the food waste receiving tanks. Manure and food waste mixture were separately fed into the test digester. Manure to food waste ratio of 0.9:0.1 was implemented at the beginning of the project. Following initial testing period, the manure to food waste ratio was maintained at 0.8:0.2 throughout the pilot testing. Organic loading rate was gradually increased from 1.53 to 3.10 kg vs per m^3-day (0.094 to 0.19 lb VS per cf-day). SRT of the system was kept between 15 and 20 days.
12	Risks and dependencies: risk assessment of project and changes	The project had to overcome the following challenges: **Community Acceptance**: Information not available. However, offensive odor associated with manure and food waste storage and operation may impose an aesthetic issue and may be inacceptable by the surrounding community. **Environmental Impact of Co-Digestion:** Production of biogas, a renewable energy that reduced the reliance on non-renewable fossil fuel power and associated pollution that occurs during fossil fuel power generation. Reduction in GHG emission caused by capture of CH_4 and NO_X which would have been emitted by the manure, that are instead collected. With co-digestion, the manure was dry-collected from the corrals within 24 hours of excretion and transported to the RP-1. This practice resulted in reduced nitrate loads to the groundwater and reduced need for reverse osmosis use for groundwater treatment. Environmental regulations are becoming more stringent, requiring improved manure collection and management. The compliance costs may adversely affect the smaller dairy's economic viability; there is a strong trend toward larger and larger diary operations. **Design and Operational Challenges:** Lack of design experience; accurate projection of gas production and gas composition is a complex task which may lead either oversized (cost penalty) or under-sized gas collection and gas treatment facilities (reduction in revenue). Many suppliers are available for competitive equipment procurement (i.e., manure storage and feed system, gas cleaning and storage, etc.). Food waste input into the digesters needs to be metered at a steady rate. "Slugfeeding" large amounts of food waste at RP-1 caused immediate spikes of gas production which can overwhelm existing gas systems. Associated with this, the food waste receiving and holding equipment should be sized carefully to be able to receive deliveries and hold enough food waste to meter it in gradually to the digesters. Ensuring that the digester is designed for appropriate loads and is not overloaded. Also ensuring that dilution is appropriate to allow good mixing in digesters. **Construction Challenges**: Low to Moderate. Piping modification to connect the modified digesters with the existing gas distribution system

Table 3-2. Co-Digestion of Dairy Manure with WWTF Sludge Case Study: IEUA RP-1 Co-Digestion of Dairy Manure to Energy, City of Ontario, California, U.S., continued.

Ref	Enquiry Item	Response Information, Description and Remarks
13	Implementation: design, build, procurement, installation and commissioning:	Design and construction services were provided separately. Operation and maintenance was provided by IEUA staff. Procurement was very similar to other wastewater projects. However, the biggest challenge was to find reliable quantity and quality of food waste sources for the co-digestion and secure them for long-term operation.
14	Energy generation: kWh & kWh/m^3	During pilot testing, manure co-digestion increased gas production by an average 620 m^3/day or 3,000 mmBtu/year compared to manure digestion only. The additional power generation with co-digestion using extra biogas was 180,000 kWh/year.
15	Cost/benefit analysis: financial appraisal or payback time	Total capital investment for manure digestion only was USD$1,100,000. Co-digestion upgrades to manure digestion were USD$268,000. Co-digestion O&M cost was USD$29,000 per year. A net increase of USD$9,000 per year over manure digestion only. Co-digestion receives approximately USD$9,000 per year renewable energy credit and USD$177,000 GHG emissions reduction credits. Life-cycle analysis indicated a rate of return between 9.7 and 9.7% while showing a simple payback period of 7.7 years for co-digestion.
16	Project review: could it be improved or developed?	The project and business model can be adopted by other utilities in the wastewater industry where additional digester capacity is available for co-digestion and manure and/or food waste sources are available. Acid-gas phase digestion as provided in IEUA RP-1 can enhance digestion rates and open up additional capacity for co-digestion. The facilities which do not have phase digestion need to evaluate their capacity before moving forward. Anaerobic digestion installations indicate that centralized anaerobic digestion installations are more likely to be economically viable. U.S. dairy farmers face a challenging business environment because of low milk prices and increasing environmental regulation. Individual farmers may have a difficult time making large long-term investments in waste treatment facilities. For most individual dairies, it is uncertain how development pressures will affect their long-term future. A centralized approach also allows for expert operation and maintenance of the facility with an emphasis on stable operation and biogas production and allows the dairy farmer to focus on operating a dairy. It also disperses the risk associated with the financial integrity of individual dairies; long-term capital investment can be made independent of the fortunes of a particular dairy. Green projects may be more attractive in near future when independent traders can purchase GHG credits even with renewable energy generated from the project being claimed in another sale.
17	Confidence grade: on data provided	The fact sheet presented above is based on information published by California Energy Commission reports and communications with IEUA staff.

3.3.2.1 Observations

Project Location

The co-digestion project is located in IEUA Regional Plant 1 (RP-1) Co-digestion of Dairy Manure to Energy, City of Ontario, California. The RP-1 is located in an industrial urban area where large dairy farms are within close proximity. Figure 3-5 shows the pilot co-digestion facilities.

Figure 3-5. Food Waste Receiving Tanks, Test Digester (Digester 4), and Piping Modifications for Pilot Co-digestion Project In IEUA RP-1.

Reprinted with permission from CH2M HILL.

3.3.2.2 Key Co-Digestion Components:

Acid Manure Digester (AMD) and Manure Digester (Digester 4): AMD has two compartments, each with a volume of 75,000 gallons, and is operated in plug-flow mode without mixing. AMD was operated without temperature control and, therefore, the temperature maintained in AMD varied as function of the ambient temperature. Manure was transferred from AMD to Digester No. 4; a completely mixed reactor operated at an average hydraulic residence time value of 19 days.

Holding Tanks and Feed System for Co-Digestion Food Waste: Four holding tanks and transfer pumps that fed food waste into the Digester 4, as shown in Figure 3-5.

Biological Gas Treatment System: The pilot system for biological treatment to remove H_2S from biogas was located in an available space to the north of Digester 4. This equipment treated biogas from Digester 4.

Power Generation Equipment: Up to 500 kW of new power generation equipment was installed to convert extra biogas from co-digestion into power.

3.4 Sludge Reduction Technologies – Focused Electrical Pulse

3.4.1 Technical Overview

3.4.1.1 Process Description

- WWTFs produce a significant volume of sludge as by-products of their water reclamation processes. These solid materials represent a large store of renewable energy, and their handling and disposal account for up to 50% of the operational expense in a modern plant.
- For WWTFs that have high nitrate concentration and insufficient influent organic carbon in the form of BOD, an external electron donor must be added to match contaminant levels (De Lucas et al., 2005). Current treatment technologies typically rely on external organic donors, most often bulk commodity chemicals such as methanol or ethanol, to drive the denitrification reactions. In plants that utilize a chemical supplemental organic electron donor to drive denitrification, the expense to import, store, and safely handle the supplemental carbon can be a major operating cost. An effective means to utilize residual sludges from the aeration process to drive denitrification would improve the plant economics for biosolids disposal and denitrification at the same time. Although components of primary and/or WAS have been studied as potential replacements for imported organic donors, they have inherent kinetic limitations (Rittmann and McCarty, 2001).
- In larger municipalities in the U.S. and other developed nations, anaerobic digestion of organic solids generated in and collected from other parts of the wastewater treatment process reduces the volume of solids generated and captures a portion of the available energy in a usable form. Anaerobic digestion traces its roots to the 1850s and is one of the oldest forms of wastewater treatment (Tchobanoglous and Burton, 1991). In the U.S., approximately 25% of the nearly 17,000 WWTFs have anaerobic digestion capabilities (U.S. EPA, 2003).The primary purpose of anaerobic digestion in these plants is the stabilization of organic matter to CH_4 and CO_2, with concurrent reductions in odors, pathogens, and solid material requiring disposal (Parkin and Owen, 1986; Speece, 1996; Rittmann and McCarty, 2001).
- Anaerobic digestion, like other biological water and wastewater treatment processes, is catalyzed by the activity of microorganisms (consult Parkin and Owen, 1986; Speece, 1996; and Rittmann and McCarty, 2001 for a detailed discussion).Unlike aerobic biological reactions, which can be performed by a single species of bacteria, the anaerobic digestion process requires a complex consortium of microbial species, including facultative and obligate anaerobes, each performing a distinct role in the overall conversion of organic material to CH_4 and CO_2.
- Parkin and Owen (1986) and Speece (1996) describe three primary phases of anaerobic digestion of wastewater sludges.
  - In the first phase, insoluble and complex organic components are hydrolyzed and converted to soluble forms suitable for use as substrates in later microbial reactions, similar to the digestion mechanism in humans and mammals that converts solid foods

into soluble compounds for absorption into the body. Until this conversion is complete, the organic compounds undergoing hydrolysis are "locked" in place, unavailable for subsequent utilization by other microorganisms; thus, this phase often is rate-limiting when the input organic material is a solid, as is the case with wastewater sludge. The hydrolysis of WAS is much slower than for the organic solids in primary sludge (Gossett and Belser, 1982; Parkin and Owen, 1986; Rittmann and McCarty, 2001), which is a primary reason why WAS is generally considered "half as digestible as primary sludge" (Zack and Edwards, 1929).

  - In the second phase of digestion, the simple sugars and organic acids are converted to even simpler constituents, mainly acetic acid, CO_2, and hydrogen (H_2). Most of the energy value of the original material is retained in the acetic acid and hydrogen.
  - Finally, in the third phase the primary end-products of Phase 2 are converted to CH_4 by a unique class of microorganisms, the methanogens. One group of methanogens uses hydrogen and carbon dioxide to produce CH_4; a second group catalyzes the cleavage of acetate to produce CH_4 and CO_2.

- During anaerobic digestion, almost all the energy value of the released from the original organic material is conserved in CH_4, which is a relatively low-solubility gas that naturally evolves from the water. The gas that evolves from the digester is in a ratio of 6 to 7 parts CH_4 for every 3 to 4 parts of CO_2. This mixture, commonly referred to as biogas, is slightly lower in energy value than natural gas and has an economic value of USD$0.005 to USD$0.01 per cubic foot at current natural gas prices.
- More than 30 years ago, it was recognized that the Phase 1 hydrolysis step is the key limit on the rate of digestion of WAS and that this limitation could be removed by "pre-hydrolyzing" the material (Haug, 1977; Gossett et al., 1978; Haug et al., 1978; Stuckey and McCarty, 1978; Owen and McCarty, 1979; Gossett and Belser, 1982; Stuckey and McCarty, 1984). The main requirement is to breach the membranes of cellular material produced in the activated sludge process. For obvious reasons, cell membranes are recalcitrant to degradation by other organisms. Cellular membranes have been described as "thick and leathery, like the shell of a turtle, that needs to be ripped open to improve the digestibility of WAS" (Speece, 2007). In addition, the cell wall provides another protective barrier that needs to be breeched to make biological solids as bioavailable as possible, as it provides structural support to the cell membrane.
- Pioneering work in the laboratory of Dr. Perry McCarty investigated the effects of thermal and chemical pre-treatment on improving the digestibility of organic sludges (Haug, 1977; Haug et al., 1978; Gossett et al., 1978; Stuckey and McCarty, 1978; Owen and McCarty, 1979; Stuckey and McCarty, 1984).This early work still provides one of the most comprehensive investigations of using sludge pre-treatment to improve the digestibility of WAS and mixed sludges; it is a model for evaluating other technologies. Haug et al., (1978) performed a series of studies on the impact of heat pre-treatment in the range of 150 to 225°C on the performance of lab-scale anaerobic digesters operated at a 15-day detention time and fed WAS or mixtures of WAS and primary sludge. These researchers demonstrated several important benefits from treating 100% WAS at the optimum treatment temperature of 175°C:
  - Increase in the soluble COD concentration by over 700%
  - Increase in volatile solids destruction rate by over 75%
  - Increase in biogas production by over 70%
  - Decrease in solids waste disposal by over 35%

- Given the energy consumption and other disadvantages of thermal pretreatment technologies, considerable research has focused on the development of alternative methods to obtain the benefits of pre-treatment on improving the digestibility of WAS and other sludges. Several reviews have compared the relative effectiveness of various technologies at the lab, pilot, and full scale (Mueller, 2000; Muller 2001; Stensel et al., 2001; Kim et al., 2002; Lafitte-Trouque and Forster, 2002; Barjenbruch and Kopplow, 2003; Kim et al., 2003; Wei et al., 2003; Muller et al., 2004; U.S. EPA, 2006; Bougrier et al., 2007a; Roxburgh et al., 2006; Khanal et al., 2007).
- Pulsed electric field (PEF) technology directly attacks the basic building blocks of all cell membranes and walls: the phospholipids and the peptidoglycan, respectively. Both are polar molecules. Both have ligand groups exposed to the environment, giving a net negative charge on the cell's outer surface (Stumm and Morgan, 1996; Madigan et al., 2003).Because of the charged and polar nature of the building blocks of cell membranes and walls, they are susceptible to the action of strong electrical fields.
- Molecular biologists exploit this susceptibility by using PEFs for electroporation, the reversible opening of pores in cellular membranes to perform plasmid and DNA insertions and for medical therapies (Aly et al., 2001; Madigan et al., 2003; Xu and Xiao, 2006).Food biologists have studied PEF treatment as an alternative to traditional pasteurization technologies for decades (Zhang et al., 1995a; Töpfl, 2006 and references therein; Töpfl et al., 2007; Zhang, 2007).Treatment using PEF has been shown to inactivate a variety of microorganisms (Mizuno et al., 1989; Zhang et al., 1994; Zhang et al., 1995b; Schoenbach et al., 1996; Schoenbach et al., 1997; Schoenbach et al., 2000; Vernhes et al., 2002; Beveridge et al., 2003; Kunitomo and Obo, 2003; Sepulveda et al., 2006; Töpfl, 2006 and references therein; Zhang, 2007) and viruses (Mizuno et al., 1990) in liquids.
- Like electroporation, when used as a sterilization technique, PEF induces the formation and opening of pores in cellular membranes. However, unlike the application of PEF for electroporation, the pores that opens in the pulsing electric field when higher electrical power is applied do not close in a reversible fashion. Instead, treatment is applied until the cell membranes become permeable to the influx of small molecules from carrier medium, leading to swelling and rupture of the cell (Mizuno, 1989; Schoenbach et al., 1995; Schoenbach et al., 2000; Loeffler et al., 2001; Fang et al., 2006; Timoshkin et al., 2006; Töpfl, 2006 and references therein; Koners et al., 2007; Sato, 2007).
- Because PEF acts directly on cellular membranes, researchers in Europe, the U.S., and Asia have recognized the potential energy advantage over other technologies for sludge pretreatment (Koners et al., 2004; Kopplow et al., 2004; Choi et al., 2006; Koners et al., 2006a; Koners et al., 2006b; Banaszak et al., 2007; Banaszak et al., 2008; Rittmann et al., 2008; Banaszak et al., 2009b; Salerno et al., 2009).In most cases investigated to date, PEF treatment of sludge has increased COD solubilization, reduced solids from digestion by 50% or more, and increased biogas production by over 50% (Koners et al., 2004; Choi et al., 2006; Banaszak et al., 2007; Banaszak et al., 2008; Rittmann et al., 2008; Banaszak et al., 2009b; Salerno et al., 2009).
- Pencil's focused pulsed (FP) technology is the full-scale application of PEF technology to make the organic material in WAS more bioavailable through the mechanism of electroporation. Once inactivated and partially solubilized by FP treatment, the organic material in WAS is made more bioavailable for downstream processes. When FP-treated material is sent to anaerobic digestion, it is more completely stabilized to CH_4 and CO_2;

when it material is sent to a denitrification process, it can be used to replace external supplemental carbon additions.

- A commercial FP unit was installed and began operations at the Mesa Northwest Water Reclamation Plant (NWWRP) in April of 2007.All material treated from September of 2007 through the end of 2008 was fed to the anaerobic digesters. Prior to September and after the end of 2008, some FP-treated material was diverted to other processes.
- Starting in April of 2009, a portion of the FP-treated material was diverted to the plant headworks to replace methanol and/or glycerol additions to drive denitrification. Supplemental methanol additions were eliminated completely and replaced by internal carbon generated by FP treatment in June of 2009 (Banaszak et al., 2009a).

3.4.1.2 Potential Benefits

- FP pretreatment of solid sludges offers significant benefits over traditional anaerobic digestion processes:
  - Improvement on digestibility and biogas production
  - Reduced biosolids handling and disposal fees
  - Reduced dewatering polymer requirements
  - Production of biogas, a renewable energy source that reduces the reliance on non-renewable forms of energy
  - Generation of all or part of the heat required to bring sludge to the appropriate temperature for anaerobic digestion
  - Reduction in GHG caused by enhanced capture of CH_4 and replacement of fossil-fuel energy and external carbon sources
  - 3 to 6x ratio of additional energy produced as heat and biogas relative to the electrical energy consumed for treatment
  - Reduced odor and pathogens in the final biosolids product
  - Potential increase in digester capacity
  - Potential reduction in micro-constituent organic compounds
  - Reduced cost and safety concerns associated with external carbon sources

3.4.1.3 Concerns

- Offsetting the benefits mentioned above, there are few potential concerns associated with the implementation and operation of a pretreatment system for enhanced anaerobic digestion:
  - The generation of additional biogas could drive the requirement to install new or upgrade existing facilities for gas cleaning, handling, and utilization.
  - Digester recycle streams returned to the head of the WWTF contain higher concentrations of nutrients (i.e. nitrogen and phosphorus) that may require further treatment.
  - For plants with multi-year biosolids disposal contracts, reduced solids mass might temporarily trigger minimum use clauses.

3.4.1.4 Range of Potential Cost/Savings

- The cost savings available from installing FP technology at any given plant depend on the size of the facility; the costs of biosolids handling, dewatering, and disposal; beneficial reuse options for the additional biogas generated; electrical and natural gas energy costs; and the potential for replacement of an external carbon source. For most facilities, the economic value of increased biogas production and reduced biosolids hauling and disposal outweighs the energy and other operating requirements of FP pre-treatment by a considerable margin.

Typically, FP operating and maintenance costs range between USD$20 and USD$30/dry ton treated.

- The energy consumption required for FP treatment, about 300 kWh/dry ton treated, is offset by the production of additional biogas. In addition, approximately 90% of the energy used for treatment is recovered as a temperature increase in the treated sludge leaving the FP unit, potentially reducing the requirement to provide an external heat source. For any installation, the true energy recovery depends on the actual reduction in external heating requirements, which under most conditions depends on the amount of heat lost between the outlet of the FP unit and the digester.
- The following table provides an example benefit calculation for a 20 mgd (75,700 m^3/day) plant treating an average of 100,000 gal/day (380 m^3/day) of WAS and primary sludge, generating an average of 215,000 scf/day (6,100 m^3/day) of biogas with an economic value of USD$0.008/scf (USD$0.28/m^3), and disposing of 13.6 dry tons/day (12,300 kg/day) of biosolids at a cost of USD$250/dry ton (USD$0.28/kg).Without accounting for any other plant benefits (e.g., reduced polymer, heat recovery, etc.), FP treatment would generate an economic benefit of approximately USD$680,000 annually net of electricity and other operating and maintenance costs when the biogas increase and biosolids decrease are 60% and 40%, respectively. For a plant this size, and depending on operating conditions, replacement of an external carbon source for denitrification could generate an equivalent amount of savings. The capital costs of the FP equipment required to treat this volume of material are approximately USD$2 million to USD$2.5 million. Typical simple capital payback calculations for a number of plants indicate that expenditures could be recaptured in one to five years.

Cost driver	Before FP	After FP	Change	Unit cost	Annual benefit
Biogas (scf/day)	215,000	344,000	129,000	$0.008/scf	$376,680
Biosolids (DT/day)	13.6	8.2	5.4	$250/DT	$492,750
Electricity, O & M				$520/day	($189,800)
Total benefit					**$679,630**

3.4.1.5 Application Potential

- FP pretreatment for improving anaerobic digestion is attractive for WWTFs receiving municipal influent of 2-5 mgd and above. Smaller plants may benefit if the influent and/or digester feed contains a significant fraction of high-strength organic material. In addition, the economics for external carbon replacement to drive biological nutrient removal processes are typically more favorable, so plants as small as 1 mgd could benefit. FP pretreatment might also be economically feasible for solids reduction at aerobic facilities with high biosolids disposal costs and excess aeration capacity.
- FP pretreatment shows significant benefits for improving the digestibility of agricultural and high-strength organic wastes. Also, FP treatment may be used to render the anaerobic biomass grown in digesters receiving large quantities of soluble organic material more amenable to digestion.

3.4.2 Sludge Reduction Technologies – Focused Electrical Pulse Case Study, OpenCEL FP Unit at the Northwest Water Reclamation Plant, Mesa, Arizona, U.S.

Table 3-3. Sludge Reduction Technologies – Focused Electrical Pulse Case Study, OpenCEL FP Unit at the NWWRP, Mesa, Arizona, U.S.

Ref	Enquiry Item	Response Information, Description and Remarks
1	Location: Country, urban or rural	Urban area
2	Sector: clean, waste or sludge	Wastewater sludge
3	Works Owner or Operator: with financial set-up, regulatory or not	**Owner and Operator:** WWTF facilities are owned and operated by The City of Mesa, AZ. The FP system is owned and operated by OpenCEL. Organization set-up: Regulated public agency. Source of revenue: Revenue is realized by charging user fees for wastewater collection and treatment services provided.
4	Size: flows and loads or population equivalent	Permitted Capacity is 68,000 m^3/d (18 mgd) Average Daily Flow is 39,000 m^3/d (10 mgd) Average Daily Loads: BOD: 12,650 kg/d TSS:16,300 kg/d TKN: 2,100 kg/d
5	Energy Provider: with costs, incentives, taxes and conditions	Electricity provided by Salt River Project at an average cost of USD$0.075/kWh in 2008; Peak summer electric rates at high as USD$0.14/kWh Natural gas provided by the City of Mesa at an average cost of USD$1.20/therm in 2008
6	Process: physical, chemical, or biological description	The OpenCEL unit is installed in the existing biosolids building for primary and WAS pretreatment prior to anaerobic digestion
7	Component: all or part of the works	Preliminary and Primary Treatment – screening, grit removal, primary settling. Secondary Treatment – traditional nitrification/denitrification activated sludge followed by secondary clarification. Tertiary Treatment – Granular media filtration and UV disinfection Solids Processing – Primary sludge and WAS thickening via centrifuge. Thickened sludge is mixed and co-digested in two egg-shaped, mesophilic anaerobic digesters. Dewatered sludge via centrifuge is hauled to land application by Solids Solutions. The biogas is conditioned and used in a CAT 550 kW cogeneration system for electricity generation.
8	Specific energy problem: including quality or consent details	Focus on increased biogas production for electricity cogeneration during peak demand periods.

Table 3-3. Sludge Reduction Technologies – Focused Electrical Pulse Case Study, OpenCEL FP Unit at the NWWRP, Mesa, Arizona, U.S., continued.

Ref	Enquiry Item	Response Information, Description and Remarks
9	Process/Plant changes: mechanical, electrical or controls	Sludge heated about 20°F during treatment and viscosity reduced. No change to plant effluent.
10	Civil/Physical changes to water/effluent quality, civil works, or process	Installation of an OpenCEL Focused Pulsed (FP) pretreatment unit, piping, grinding/pumping equipment, and control integration. No other changes to the plant.
11	Operational Changes: skill levels, procedures and maintenance routines	Optimization of cogeneration operation to take advantage of additional biogas. Leveling of thickened sludge flow to better optimize the capacity of the FP unit. FP treatment chamber changed periodically by vender (20-minute activity).
12	Risks and Dependencies: risk assessment of project and changes	-
13	Implementation: design, build, procurement, installation and commissioning	Solids handling suspended for less than one day to accommodate installation of FP equipment. OpenCEL maintains ownership of FP equipment and is compensated via a savings-sharing agreement.
14	Energy Efficiency gains: kWh & kWh/m^3	1,000 to 1,150 m^3/day (35,000 – 45,000 scf/day) more biogas is produced by the digesters with FP pretreatment. On average for 2008, 4.8 times more energy was produced as heat and additional biogas than was consumed for FP treatment.
15	Cost/Benefit analysis: financial appraisal or payback time	The maximum level of FP treatment obtained (3-month moving average of treated volume to the digesters) peaked at 71% in July of 2008 and dropped off during installation of a closed-loop cooling system during the summer of 2008; the OpenCEL unit treated 53% of the total sludge volume fed to the digesters in 2008. On a sludge-flow-normalized basis, biosolids trucked from the NWWRP in 2008 decreased by 17% as compared to a prior 3-year baseline; the projected decrease at full treatment level is 30%. On a sludge-flow-normalized basis, biogas production at the NWWRP in 2008 increased by 32% as compared to a prior 14-month baseline; the projected increase at full treatment level is 58%. On a sludge-flow-normalized basis, natural gas consumption at the NWWRP decreased by 58% in 2008 compared to 2006; natural gas consumption is expected to decrease further with additional treatment. The City of Mesa is saving up to USD$9,000 per month net of FP operating costs from elimination of methanol additions, cogeneration of additional biogas into electricity, reduced natural gas consumption, and reduced biosolids to disposal.

Table 3-3. Sludge Reduction Technologies – Focused Electrical Pulse Case Study, OpenCEL FP Unit at the NWWRP, Mesa, Arizona, U.S., continued.

Ref	Enquiry Item	Response information, description and remarks
16	Project review: could it be improved or developed?	Level of savings will increase with treatment of a higher percentage of plant sludge flow. Green projects may be more attractive in near future when independent traders can purchase GHG credits in a robust market.
17	Confidence grade: on data provided	The technology described is highly transportable and adaptable. On a scale of 1 through 5, the confidence grade provided to the information presented above is 5.

3.4.2.1 Observations

Project Location

The location of this study was the NWWRP in the City of Mesa, Az. The NWWRP plant is a state-of-the-art wastewater reclamation facility, with a maximum treatment capacity of 18 million gallons per day, and treats an average of 10 mgd of wastewater daily. This facility has unit processes that include screening, grinding, sedimentation, organics removal, anaerobic digestion, nutrient removal, filtration, clarification, and disinfection. The plant maintains an aerobic sludge age between 10-15 days to achieve full nitrification/denitrification and has an effluent discharge limit of 10 mg/L for total nitrogen. The effluent from the NWWRP is discharged to two recharge sites and the Salt River, which also recharges the aquifer. Solids handling processes at the NWWRP facility include sludge thickening, which produces thickened primary and WAS at the rate of approximately 50,000 gallons per day, two anaerobic digesters, and a centrifuge dewatering installation. Biosolids from the centrifuge are trucked to land application. The plant also has an electric power co-generation unit that is capable of using biogas produced by the anaerobic digesters for generating electricity. Figure 3-6 is a schematic diagram showing the location of the OpenCEL FP unit, which was installed between the sludge thickeners and digesters and is capable of treating a majority of the flow to the digesters.

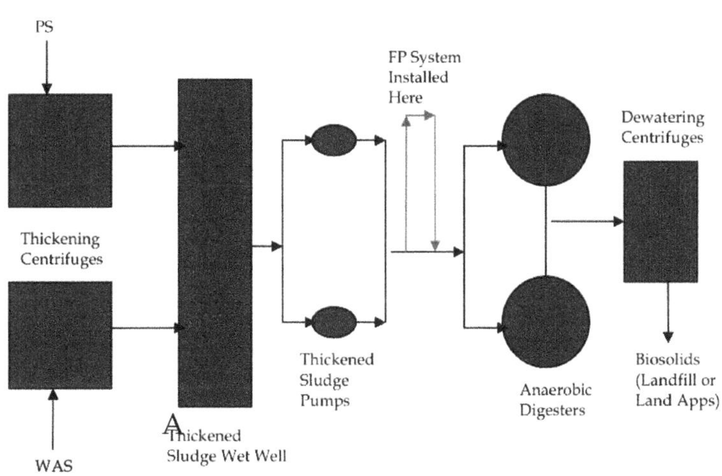

Figure 3-6. Process Flow Diagram for the Mesa NWWRP Showing the Installation Point for the FP Unit. Reprinted with permission from OpenCEL.

3.4.2.2 OpenCEL FP Components:

OpenCEL initiated a commercial, full-scale installation of the FP technology at the NWWRP in March of 2007. An OpenCEL FP unit was installed at the facility to treat the total flow of primary and WAS before entering the plant's two anaerobic digesters (Banaszak et al., 2007; Banaszak et al., 2008; Rittmann et al.., 2008; Banaszak et al., 2009b). The design treatment capacity of the unit is approximately 50,000 gallons per day (35 gpm or 7,950 lph) at a 5-6% total solids content. Auxiliary equipment includes piping and valving, cooling and flush water supply, a pump and grinder, various metering devices, and a process control system with remote control capabilities. Installation of the unit was completed over a several-week period with minimal disruption to plant operations. The FP unit consisted of the aforementioned control system, a high-voltage power supply, and a modulator unit with attached flow-through treatment chambers. Sampling capabilities were provided before and after the FP treatment chamber. Figure 3-7 shows the major components of the OpenCEL FP system.

The full-scale FP unit began start-up operations at the NWWRP plant in March of 2007.After a several month shakedown period and apart from regular maintenance and approximately one month of down time for the addition of cooling equipment in July/August 2008, the FP unit has been operating continuously and unattended. After the cooling modifications were completed in August 2008, plant staff requested that OpenCEL run the unit at partial capacity to avoid exceeding the digester over-temperature alarms (set at 100°F; to be modified before summer 2009) during the warm summer/fall months. As of November 2008, plant personnel removed the treatment constraints.

In April 2009, permanent full-scale denitrification began at Mesa NWWRP using FP-treated material fed through an automated plant mechanical system. Plant SCADA results indicated that FP-treated sludge is an effective electron donor for denitrification when injected at the plant head works and passed through the primary clarifiers before entering the anoxic zones. Figure 3-8 summarizes the volume of FP-treated material and the secondary clarifier effluent nitrate concentrations for the months of March through July, 2009. Nitrate concentrations dropped significantly after the FP-treated material was diverted to the plant head works, and start-up issues were resolved during the first week of treatment (week ending 4/18/09).Specifically, nitrate concentrations dropped by nearly 50% compared to the prior two-week average during the week of April 20, the first week of full-scale operation with both external and internal carbon additions. Plant staff slowly reduced external carbon additions through the summer and slightly increased the volume of FP material diverted to the headworks. External carbon feed was suspended completely after June 20, 2009, and was replaced by approximately 2,200 gallons of FP-treated material daily. The plant operated steadily from late June through July with no significant impact on effluent nitrate (or total N) concentrations.

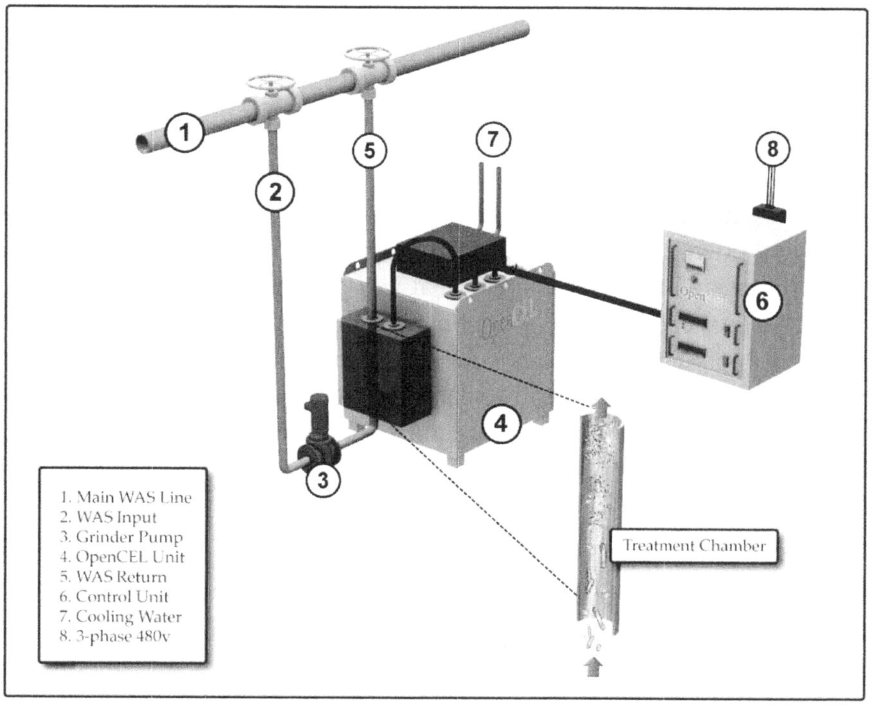

Figure 3-7. Major Components of the OpenCEL FP System.
Reprinted with permission from OpenCEL.

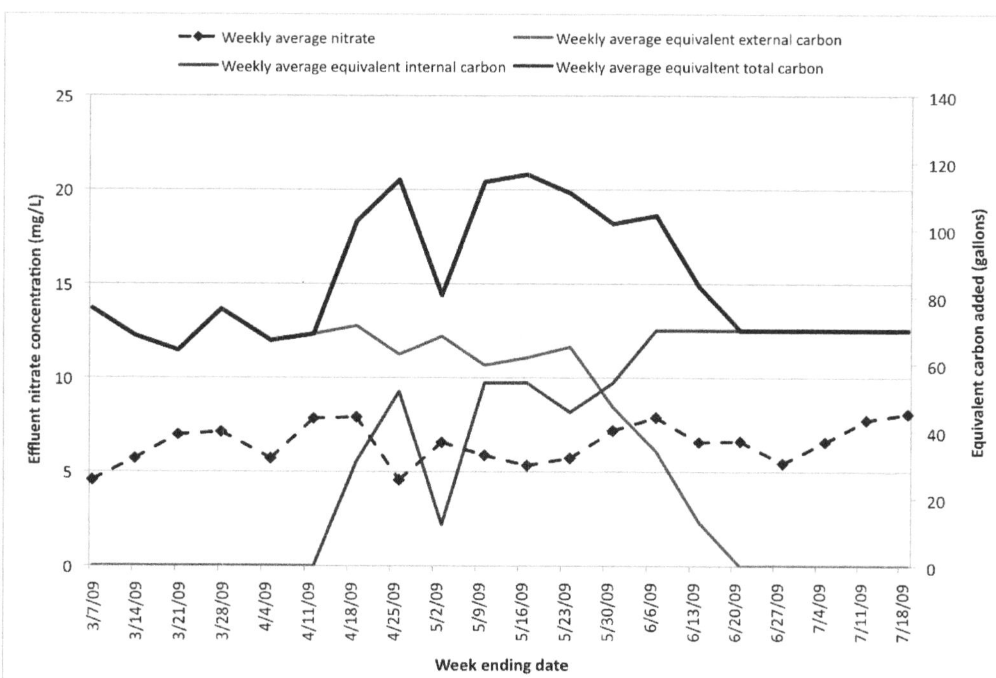

Figure 3-8.. Summary of Carbon Additions and Daily SCADA Trends from the Mesa NWWRP after Initiation of Full-Scale Operations Using FP-treated Materials.
Reprinted with permission from OpenCEL.

3.5 Biogas Cleaning Technologies – Siloxane Removal

3.5.1 Technology Overview

3.5.1.1 Process Description

With the growing cost of energy and the concern of GHG emissions, municipal wastewater treatment plants (WWTFs) are looking to utilize digester gas as a fuel source for production of heat and power generation. In recent years a number of plants have started utilizing digester gas in cogeneration of electricity and heat using gas engines/generators. In the early days of cogeneration, it was recognized that contaminants such as water and H_2S could cause damage to engines through acid corrosion to the engine parts and indirectly through contamination of the lubricating oil. Engine manufacturers included limits for H_2S in their engine specifications. More recently, they have included limits on moisture and other contaminants, such as siloxanes in their engine specifications.

Siloxanes enter the sewer and the WWTP in the liquid phase and are transferred to the digesters with the sludges. Within the digestion process, the siloxanes change from the liquid phase into the gaseous phase within the digester gas. Silicate and silica are formed during the combustion of digester biogas containing siloxanes. Siloxanes combine with free oxygen or other elements in combustion gas using the principles of thermodynamics, combustion and pressure. Build up within the system can lead to increased wear on the engine, increased levels of silicon within the lubrication oils as well as clogging and improper sealing of the valves. Silica deposits also decrease the transfer efficiency of heat exchangers. Siloxane removal has been successfully implemented for a number of years. The most common removal process is a combination of refrigerated gas drying along with spherical artificial graphite (SAG) media adsorption (Burrowes et al., 2005).

Refrigeration/condensation has been used effectively to remove siloxanes from digester gas in a number of facilities. This process removes compounds by sending the gas through a refrigerant system, a moisture separator and a reheat exchanger. The condensate from this system is usually recycled back to the plant. These systems are effective by themselves only when the gas is compressed to medium to high pressures (700-3,500 kPa).

SAG graphite mol sieve media is commonly used to remove siloxane from digester gas. Siloxane is removed by mass transfer via solid phase adsorption. The SAG vessels may be installed in series or in parallel, however most units are installed in series. This configuration allows the first vessel to remove most of the siloxane with a second unit as a polisher. By pass valves and piping allow one vessel to be taken offline to replace the media as required. The vessels can be installed indoors or outdoors. They typically consist of silo type tanks with valved inlet and outlet manifolds that are arranged to feed digester gas to the bottom of the tanks and discharge at the top. The tanks contain walkways and access hatches at the top of the tank and a coned bottom. The media is supplied in super sacks (1.14 m^3, 500-600 kg), drums (0.2 m^3, 90-110 kg) or paper sacks (0.06 m^3, 25-30 kg).

3.5.1.2 Potential Benefits

Removing siloxanes from the digester gas prior to combustion will provide the following benefits:

- General:
  - Reduction of siloxanes from the digester gas to trace level measurements

- Lower operating and maintenance costs

- Cogeneration:
  - Cleaner fuel and increased operational efficiencies
  - Longer intervals between maintenance for cogeneration equipment
  - Decreased down time for equipment
  - Longer spark plug life
  - Increased life of engine oil
  - Extending life of engine heads, cylinder linings, pistons, impellers and heat recovery components
  - Increasing engine runtime
- Boilers:
  - Longer intervals between maintenance
  - Cleaner fuel and increased operational efficiencies
  - Increased heat transfer capabilities
  - Increased life of ignition system and combustion chamber
  - Increased life of boiler tubes

3.5.1.3 Concerns

- Siloxane control systems add to the level of complexity of a facility.
- Refrigeration to 4°C only removes a fraction of the siloxanes. SAG
- Refrigeration below 4°C consumes more electricity and many systems have experienced freezing problems.
- Activated carbon or SAG media adsorption must be provided together with refrigeration/condensation for effective siloxane control
- Activated carbon and SAG media must be periodically replaced and spent material must be regenerated or discarded.

3.5.1.4 Range of Potential Costs/Savings

Costs for siloxane removal systems are highly specific to the actual technology employed, but in general they should range in the order of USD$ 0.02-0.03/ m^3 of biogas to be treated. This technology is aimed at improving the reliability and reducing the operation and maintenance requirements of digester biogas co-generation systems, and any potential savings are associated with this goal are by definition very particular to a specific installation.

3.5.1.5 Application Potential

Treatment plants larger than 20-40 ML/d are possible candidates for cogeneration (WEF MOP 8, 2009). Siloxane removal has recently become an issue for many of existing facilities involved in digester biogas co-generation, and any facility looking into establishing a co-generation system must consider its removal along with other contaminants such as H_2S, moisture and particulates.

3.5.2 Biogas Cleaning Technologies – Siloxane Removal System Case Study, Barrie WPCC, Ontario, Canada

Table 3-4. Biogas Cleaning Technologies - Siloxane Removal System Case Study, Barrie WPCC, Ontario, Canada.

Ref	Enquiry Item	Response Information, Description and Remarks
1	Location: Country, urban or rural:	Urban
2	Sector: clean, waste or sludge:	Sludge
3	Works Owner or Operator: with financial set-up, regulatory or not.	Owner and Operator: City of Barrie Organization set-up: Regulated public agency
4	Size: flows and loads or population equivalent:	Currently being expanded to 76 ML/d (March 2010 scheduled completion).
5	Energy Provider: with costs, incentives, taxes and conditions:	NA
6	Process: physical, chemical, or biological description:	The Barrie WPCC is configured with preliminary treatment, primary clarification, UNOX activated sludge system, tertiary treatment with RBC's, and UV disinfection. Residual sludges are anaerobically digested and biogas is used for co-generation. Biogas treatment consists on siloxane removal using refrigeration followed by graphite mol sieve media (SAG). System provided by Applied Fluid Technologies.
7	Component: all or part of the works:	Siloxane removal system for biogas prior to cogeneration.
8	Specific energy problem: including quality or consent details:	Enhancement of cogeneration system installed to utilize energy from digester biogas.
9	Process/Plant changes: mechanical, electrical or controls:	Added the refrigeration and SAG vessels.
10	Civil/Physical Changes: to water/effluent quality, civil works, or process:	Improved operational and maintenance conditions for the biogas co-generation system.
11	Operational Changes: skill levels, procedures and maintenance routines:	Vendor replaces vessel media on a semiannual basis. Media change out depends on concentration of siloxane which can vary.
12	Risks and Dependencies: risk assessment of project and changes.	Relatively new technology requires operator training and adds to operational and maintenance complexity of treatment facilities.
13	Implementation: design, build, procurement, installation and commissioning:	Conventional design, bid construct.
14	Energy Efficiency gains: kWh & kWh/m^3	System upgrades were aimed at improving co-generation system reliability rather than improve energy efficiencies.
15	Cost/Benefit analysis: financial appraisal or payback time.	Intent of system was to improved reliability of cogeneration by reducing downtime and refurbishment costs.
16	Project review: could it be improved or developed?	Technology has been widely used over the last five years or so and is considered matured.
17	Confidence grade: on data provided.	High confidence level

3.5.2.1 Observations

The Barrie WPCC operates a biogas co-generation system including the siloxane removal and is shown in Figure3-9. The co-generation system consists of two boilers and two Waukesha cogeneration engine/generator systems. The engines are turbo-charged and run at an average gas flow rate of 4,000 m^3/d. The cogeneration and boiler system operates in a lead/lag standby operations mode based on varying gas pressure levels and heat demand. Each unit process is brought on line and progressively incremented based on digester gas pressure levels. Heat recovery from the cogeneration engine water jackets and exhaust is recovered through heat exchangers and provides a thermal energy supply to service base heating loads for process (digester) and building (domestic) heating for the plant via the plant heating loop. The boilers are used to supplement heat from the cogeneration system and are fired on digester gas, if available or natural gas.

The biogas pretreatment equipment was installed in 2004 and consists of a refrigerated gas dryer and a media adsorption system located downstream of the gas booster and upstream of the 0.3 micron coalescing filter. The refrigerated gas dryer includes an air-cooled refrigerant system, a moisture separator and a reheat exchanger to raise the gas temperature above its dew point. The media adsorption system consists of dual bed upflow towers arranged to operate in series. Either tower can be taken out of service for media replacement without interrupting normal operations.

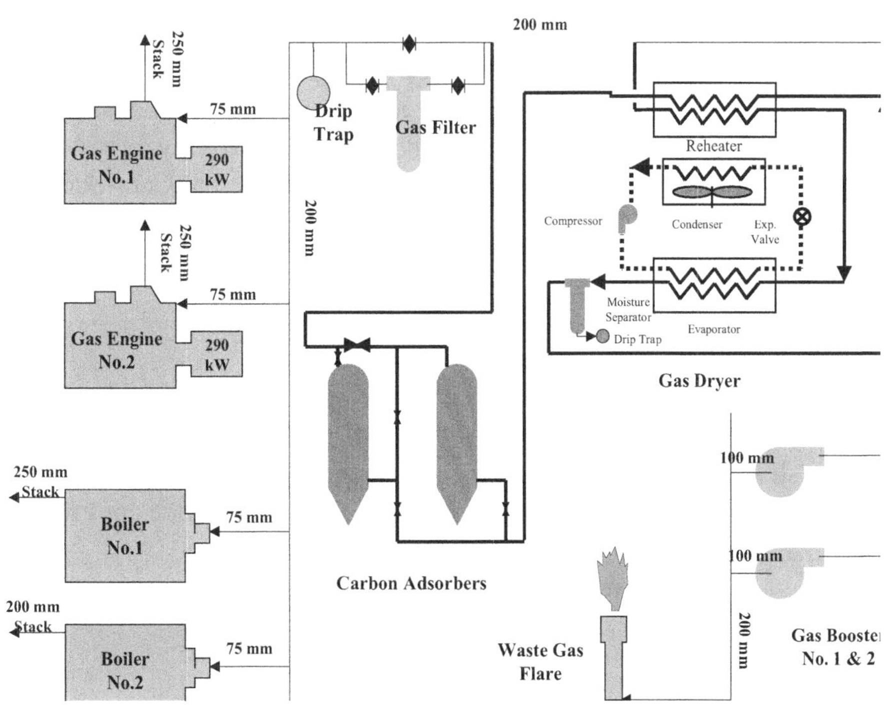

Figure 3-9. Biogas Co-generation System with Siloxane Removal, Barrie WPCC, Canada.

Reprinted with permission from Burrowes et al, 2005.

3.6 Anaerobic Digester Mixing – Linear Motion Mixers

3.6.1 Technology Overview

3.6.1.1 Process Description

- Wastewater treatment facilities routinely use digestion for the treatment of solids. The process is used to stabilize sludge, inactivate pathogens, produce biogas (a recoverable resource) and reduce odor. Anaerobic digesters are typically temperature-controlled to grow mesophilic organisms.
- Mixing of the anaerobic digester is required to:
  - Maximize contact time between active biomass and substrate
  - Create homogenous chemical and temperature conditions
  - Maximize effective volume
  - Dilute any toxic influent quickly
  - Minimize foam and scum production
  - Prevent settling of inorganic material
- There are three types of mixing commonly found in anaerobic digesters:
  - Mechanical – These mixers typically use impellors, propellers, or turbine wheels.
  - Pumped – Digester contents are recycled using external pumps.
  - Gas recirculation – Digester gas is recirculated through diffusers or internal and external draft tubes.
- Mixer performance is often characterized through chemical tracer tests or temperature profiles throughout a reactor. There is no standard mixing value, but traditional recommendations range from 5.2 to 40 W/m^3 (0.2 to 1.5 $hp/1,000\ ft^3$). This could represent approximately 10-15% the energy use in a wastewater treatment facility.
- Vertical linear motion (VLM) mixers are a new technology used to mechanically mix digesters. These mixers use a vertically oscillating disc to create eddy currents throughout the reactor with a single mixer. The mechanism is driven by an internal Cam-Scotch-Yoke system. This technology is patented by Enersave Fluid Mixers Inc., which advertises mixing with 50-90% less power than traditional methods.

3.6.1.2 Benefits

- VLM mixers use less power than other mixing methods.
- VLM mixers have fewer moving parts than most mixing technologies, which may reduce maintenance. One VLM mixer may replace several mechanical mixers.
- Changing mixers should have no effect on the treatment process if adequately sized.
- Maintenance of the mixer is primarily in the drive mechanism and does not require removal from the tank.
- Technology can be applied to fixed or floating covers.

3.6.1.3 Concerns

- VLM mixer technology is new, with few full-scale installations to date.
- VLM mixers have not been used with high solids loading in full-scale sites. Mixing performance needs to be confirmed.

3.6.1.4 Range of Potential Savings

- The potential for cost saving is site-specific. A 7.5 kW (10 hp) VLM mixer for a 5,581 m^3 (1,500,000 gallon) digester cost USD$100,000 in 2007. This makes the capital cost of VLM mixers in new digesters comparable in price to other mixing technologies.
- VLM mixers are expected to use 50-90% less power than traditional mixing technologies.
- Replacement of existing mixers may be cost-effective. Assuming the costs and efficiency gains listed above, a USD$0.10/kw*hr energy cost, and operation 95% of the time, payback may range from 2 to 16 years. This does not quantify any potential savings from reduced maintenance.

3.6.1.5 Application Potential

Digesters are commonly used in medium to large WWTFs. The most energy-intensive aspects of this process is heating and mixing. VLM may allow a large reduction in mixing energy use in all large reactors.

In thin sludge, the VLM mixers appear to have equivalent performance to other mixers. This technology has not been used at full-scale with thick sludge, and it is unknown whether performance will be consistent in those cases.

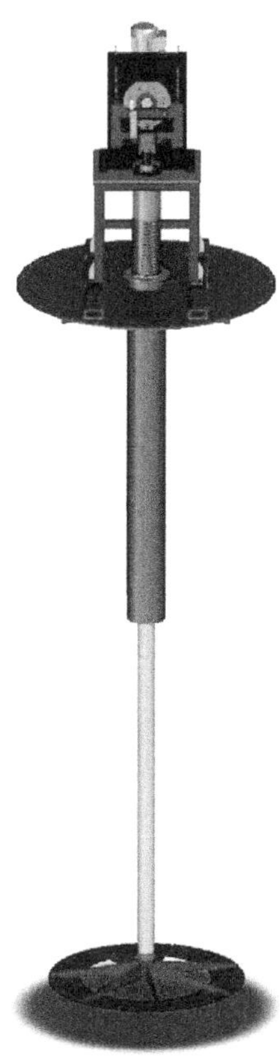

Figure 3-10. Vertical Linear Motion Mixer from Enersave Fluid Mixers Inc.

Reprinted with permission from EIMCO, 2009.

3.6.2 Anaerobic Digester Mixing – Linear Motion Mixers Case Study, Ina Road WWRF, Tucson, Arizona, U.S.

Table 3-5. Anaerobic Digester Mixing – Linear Motion Mixers Case Study, Ina Road WWRF, Tucson, Arizona, U.S.

Ref	Enquiry Item	Response Information, Description and Remarks
1	Location: Country, urban or rural	Urban
2	Sector: clean, waste or sludge	Wastewater
3	Works Owner or Operator: with financial set-up, regulatory or not	Owner and Operator: Pima County Regional Wastewater Reclamation Department (PCRWRD) Organization set-up: Regulated public agency
4	Size: flows and loads or population equivalent	Permitted Capacity: 142,000 m^3/d (37.5 mgd) Average Daily Flow: 98,000 m^3/d (26 mgd) BOD: 27,000 kg/d (60,000 ppd) TSS: 28,000 kg/d (62,000 ppd)
5	Energy Provider: with costs, incentives, taxes and conditions	Tucson Electric Power Co. USD$0.08 - $0.10 per kW*hr
6	Process: physical, chemical, or biological description	Liquid Treatment Process: Primary treatment, secondary treatment, and disinfection.
7	Component: all or part of the works	Preliminary and Primary Treatment – coarse screening, fine screening, grit removal, primary settling. Secondary Treatment – high purity oxygen basins and anoxic/aerobic basins, followed by secondary clarifiers. Tertiary Treatment – hypochlorite disinfection. Solids Processing – Primary sludge thickening via gravity thickeners. WAS thickening via dissolved air flotation. Thickened sludges go to anaerobic digestion, followed by centrifuge thickening.
8	Specific energy problem: including quality or consent details	Four 5,581 m^3 (1,500,000 gallon) anaerobic digesters were mixed using rotating impeller draft tube mixing requiring a significant input of power. Draft tube system installed power rating was 11.8 W/m^3 (0.45 hp per 1,000 ft^3) of digester volume. The digesters operate at a relatively low volatile solids loading rate of 1.3 to 1.4 kg/d*m^3 (80-90 ppd VS/kcf) and long SRT (20-30 days).
9	Process/Plant changes: mechanical, electrical or controls	Impeller draft tube mixers decommissioned on one of the anaerobic digesters with gas-holder cover. Installed one VLM mixer manufactured by Enersave Fluid Mixer, Inc.
10	Civil/Physical changes to water/effluent quality, civil works, or process	No changes.
11	Operational Changes: skill levels, procedures and maintenance routines	Reduced maintenance because of fewer moving parts compared to mechanical draft tubes. All parts requiring regular maintenance are outside the digester.

Table 3-5. Anaerobic Digester Mixing – Linear Motion Mixers Case Study, Ina Road WWRF, Tucson, Arizona, U.S., continued.

Ref	Enquiry Item	Response Information, Description and Remarks
12	Risks and Dependencies: risk assessment of project and changes	Digester Performance: Performance depends on good mixing to eliminate stratification, ensure biomass contact time, suspend grit, and prevent short circuiting. Mixing depends on the consistency of the fluid and the energy used to move it. The Ina Road WRF operated at low volatile solids loading rates of 1.3 to 1.4 kg VS/d*m^3 (80-90 ppd VS/kcf), which may require less energy for satisfactory mixing. Typical loading rates are approximately 1.6 kg/d*m^3 (100 ppd-VS/kcf). Higher solids concentration can affect fluid viscosity and thereby mixing efficiency.
13	Implementation: design, build, procurement, installation and commissioning	PCRWRD purchased and installed one VLM mixer with installation supervision by the manufacturer. Operations staff performed parallel tracer tests of the VLM-mixed digester and one digester with the old draft tube mixers. Computational Fluid Dynamics (CFD) models were run by Enersave Fluid Mixers Inc. to predict mixing, and then correlated with the tracer tests to verify the model. Tracer tests and modeling suggested equivalent mixing in both digesters.
14	Energy Efficiency gains: kWh or kWh/m3 before and after implementation	Approximately 90% savings in mixing energy. Draft tubes operate at 54 kW (73 hp) and VLMM operates at 6 kW (8 hp) per digester.
15	Cost/Benefit analysis: financial appraisal or payback time	The VLM mixer cost is USD$100,000 per digester. This results in a 2.5 year payback, assuming 48 kW (65 hp) reduction in mixer energy use, in-service 95% of the time, and USD$0.10 /kW*hr electricity rate.
16	Project review: could it be improved or developed?	Replacement of the existing mixers was accomplished with no significant problems. PCRWRD has replaced all digester mixers with VLM mixers because operators were very satisfied with its performance.
17	Confidence grade: on data provided	The fact sheet presented above is based on information published by PCWRD, communications with Ina Road WRF staff, and information supplied by Enersave Fluid Mixers Inc. On a scale of 1 through 5, the confidence grade provided to the information presented is 4.

3.6.2.1 Observations

Project Location

The Ina Road WWRF is located on the north edge of Tucson, Arizona.

3.7 Co-generation – External Combustion Engines for Anaerobic Digestion Biogas

3.7.1 Technology Overview

3.7.1.1 Process Description

- WWTFs routinely use anaerobic digestion for the treatment of solids. The process is used to stabilize sludge, inactivate pathogens, produce biogas and reduce odor. Anaerobic digesters are typically temperature-controlled to grow mesophilic organisms.
- Anaerobic digestion produces CH_4, a GHG. Methane is combustible and can be used for energy recovery. Burning CH_4 converts it to CO_2, a less potent GHG, and water vapor.
- Burning CH_4 produces thermal energy that can be wasted (flaring), used for heating (boilers), used for electricity (generation), or can simultaneously produce electricity and provide heating (cogeneration). Anaerobic digesters usually require supplemental heating, and equipment throughout the treatment plant requires electricity. Cogeneration is typically the most energy-efficient use of the CH_4 because both heat and electricity are utilized locally.
- The gas from anaerobic digesters varies in quality because of operating conditions and the composition of the digester influent. In addition to CH_4, digester gas can include carbon dioxide, hydrogen, nitrogen, hydrogen sulfide, water vapor, and other gases.
- Internal combustion engines are commonly used with digester gas and are relatively inexpensive. Any siloxane in the gas will oxidize to silicon dioxide, which may deposit on pistons and cylinder heads, damaging the engines and increasing the frequency of maintenance. Costly pretreatment of the digester gas is often required to remove siloxane and protect the engine.
- In an external combustion (Stirling) engine, the digester gas never comes into contact with moving parts and the engine operates at lower temperatures. Any silicon dioxide deposit does not damage the engine because it is not on the pistons, and the low temperature makes silicon dioxide easier to remove during regular maintenance. This eliminates most pretreatment of digester gas.

3.7.1.2 Potential Benefits

- Stirling engines are designed to run on a wide range of fuels. The only pretreatment necessary is moisture removal and compression to 13.8 kPa (2 psi). This can significantly reduce costs in some cases.
- Silicon dioxide deposits do not harm the engine and can be removed during regular maintenance. The engine uses automotive technologies, requiring minimal specialized knowledge to maintain.
- Emissions are low, and no treatment of exhaust is typically necessary.
- Stirling engines can be packaged in a modular fashion, allowing easy installation and expandability.

3.7.1.3 Concerns

- Stirling Biopower began marketing Stirling engines in 2009 and is the only manufacturer known to be specializing in digester gas applications. The client service network may need time to mature.
- The only engine currently sold by Stirling Biopower is 43 kW. The number of engines required at large WWTFs may be prohibitive.

3.7.1.4 Range of Potential Savings

- The potential savings is highly site-specific and depends on digester gas quality, available gas quantity, local electricity rates, and existing infrastructure.
- The installed cost of Stirling Biopower's engine is USD$3,000-3,500/kW. If electricity rates are USD$0.10/kW*hr, the engine runs 90% of the time, and the gas is not currently utilized, the simple payback time is 4-4.5 years. This does not consider the added value of heat recovery, or the cost of maintenance.
- Some government grants may be available for using digester gas as a renewable energy source.

3.7.1.5 Application Potential

- Anaerobic digesters are commonly used in medium to large WWTFs. All plants have electricity and heating requirements. Digester gas is frequently used for digester heating, but any excess gas is often flared because of the expense and complexity of typical internal combustion power generation equipment as well as that for gas pretreatment.
- Stirling engines may make digester gas utilization economical in cases where pretreatment was prohibitively expensive. Although they are currently more expensive than traditional internal combustion engines, the simplified maintenance and the elimination of gas pretreatment may make Stirling engines less expensive in many cases.
- Stirling engines currently marketed are 43kW. Large WWTFs may require many units. In those cases, a larger traditional engine with gas pretreatment may be preferred.
- Methane gas from multiple sources, including landfills, can be combined and used in Stirling engines.
- Use of the engine for cogeneration eliminates the need for a separate boiler to provide heat.

3.7.2 Co-generation – External Combustion Engines for Anaerobic Digestion Biogas Case Study, Corvallis WWRF Corvallis, Oregon, U.S.

Table 3-6. Co-generation – External Combustion Engines for Anaerobic Digestion Biogas Case Study, Corvallis WWRF, Corvallis, Oregon, U.S.

Ref	Enquiry Item	Response Information, Description and Remarks
1	Location: Country, urban or rural	Urban
2	Sector: clean, waste or sludge	Wastewater
3	Works Owner or Operator: with financial set-up, regulatory or not	Owner and Operator: City of Corvallis Organization set-up: Regulated public agency
4	Size: flows and loads or population equivalent	Permitted Capacity: 37,000 m^3/d (9.7 mgd)
5	Energy Provider: with costs, incentives, taxes and conditions	Pacific Corp. Industrial rate plus 10% from renewable energy. USD$0.0522/kW-hr
6	Process: physical, chemical, or biological description	Liquid Treatment Process: Primary treatment, secondary treatment, and disinfection.

Table 3-6. Co-generation – External Combustion Engines for Anaerobic Digestion Biogas Case Study, Corvallis WWRF, Corvallis, Oregon, U.S., continued.

Ref	Enquiry Item	Response Information, Description and Remarks
7	Component: all or part of the works	Preliminary and Primary Treatment –screening, grit removal, primary settling. Secondary Treatment – trickling filters and activated sludge, followed by secondary clarifiers. Tertiary Treatment – hypochlorite disinfection. Solids Processing – Primary sludge and WAS thickening via gravity thickeners. Anaerobic digestion.
8	Specific energy problem: including quality or consent details	25% of digester gas was used for heating. The remaining gas was wasted through flaring. This resource could be recovered for generation of electricity to increase plant sustainability. Agency's goal is to become 100% self sufficient in producing energy from renewable energy sources.
9	Process/Plant changes: mechanical, electrical or controls	External combustion Stirling engine installed by manufacturer as test unit. A total of three models have been tested at Corvallis. Plant operators installed water, gas, and power utilities. Added a small digester gas compressor to reach 2 psi.
10	Civil/Physical changes to water/effluent quality, civil works, or process	No changes
11	Operational Changes: skill levels, procedures and maintenance routines	All maintenance was supplied by manufacturer. Installation was 2.5 hours. Engine rebuilt in eight hours. No operational changes for Corvallis operations staff.
12	Risks and Dependencies: risk assessment of project and changes	Reliability: Stirling engines are a new product with an immature service and sales network. Utility Integration: It can be difficult to negotiate interconnect with some power companies. Emissions: Emissions quality from engine may vary from that of a flare. This should be evaluated on a case by case basis.
13	Implementation: design, build, procurement, installation and commissioning	Engines were installed as a research and development agreement with manufacturer. Corvallis spent USD$2,200 in hardware for utilities and gas compressor. Engine cost and installation of unit covered by manufacturer. Corvallis continued operation of its boiler and did not use the Stirling engine for heat recovery because engine installation was temporary.
14	Energy Efficiency gains: kWh or kWh/m3 before and after implementation	Manufacturer has tested 25 kW, 50 kW, and 43 kW engines on site. The only model currently marketed by Stirling Biopower is 43 kW. 75% efficiency with CHP generation, or 27% efficiency for power alone.
15	Cost/Benefit analysis: financial appraisal or payback time	Corvallis had negligible costs caused by research and development agreement. Engine and energy were provided free. The only pre-treatment of digester gas is moisture removal and 2 psi compression.

Table 3-6. Co-generation – External Combustion Engines for Anaerobic Digestion Biogas Case Study, Corvallis WWRF, Corvallis, Oregon, U.S., continued.

Ref	Enquiry Item	Response Information, Description and Remarks
16	Project review: could it be improved or developed?	Research continues to mitigate silicon dioxide buildup. Stirling engine went to market 2009, so service network is limited. City of Corvallis staff are very optimistic and would like to use Stirlings for all their digester gas utilization and replace the boiler in the future.
17	Confidence grade: on data provided	The fact sheet presented above is based on communications with Corvallis WRF staff, and information supplied by Stirling Biopower. On a scale of 1-5, the confidence grade provided to the information presented is 5.

3.7.2.1 Observations

Project Location

The Stirling engine was tested at the Corvallis WWRF in Corvallis, Oregon. The unit was installed outdoors. The WWRF is in a residential neighborhood.

3.8 Co-Generation – Fuel Cell

3.8.1 Fuel Cell

Process Description

- Fuel cells are devices that combine hydrogen with oxygen to continuously produce electricity by means of electrochemical reactions.
- Chemical energy is continuously converted to electricity in a non-combustion based manner by extracting hydrogen from methane in the fuel (anaerobic digester gas) delivered to the unit with ambient air providing the necessary oxygen. The disassociation of the hydrogen and oxygen, along with the formation of water, produces electricity.
- The primary components of a fuel cell are as follows:
  - A *gas cleanup unit* purifies the digester gas and removes potential contaminants. Digester gas must be cleaned prior to introduction to the gas reformer.
  - The *reformer* combusts small amounts of fuel to vaporize water and produce steam. The *reformer* then mixes this pressurized high temperature steam together with the pure methane gas from the gas cleanup module to produce the hydrogen gas essential to the fuel cell operation.
  - The *cell stack* is the electrochemical device that actually produces electricity from the hydrogen gas. The *cell stack* consists of a fuel electrode (also called the anode) where the negative charge is generated and a cathode, where the positive charge is generated by an oxidant. The electrode and cathode are separated by a thin ion-conducting membrane. The distinguishing feature amongst the various types of fuel cells is the electrolyte employed in their respective cell stacks.
  - The inverter consists of electrical devices that convert the DC electric power created by the fuel cell stack into AC and transforms this low voltage AC power into the required system voltage.

3.8.1.2 Potential Benefits

- High electric power generation efficiency.
- Extremely clean exhaust emissions.
- Heat from the electrochemical reaction is available for digester heating but rarely in sufficient amounts to supply the entire quantity of heat required by the digestion process.
- Suitable for outdoor installation and does not require a dedicated building for sound attenuation.

3.8.1.3 Concerns

- Fuel cell stacks are exceptionally sensitive to certain impurities and require exceptionally pure, clean and pressurized methane gas. If not properly conditioned, anaerobic digester gas can severely limit fuel cell operation and has been known to poison cell stacks. Even minute amounts of sulfur, as hydrogen sulfide or carbonyl sulfide, can contaminate the exotic metals in the cell stack.
- Most fuel cell based CHP systems require a supplemental heat source, often a boiler, to make up the shortfall of heat energy not provided by the fuel cells.
- In terms of the capital cost per kW produced, and when considering operating costs and cell stack replacement costs, fuel cells are one of the most expensive CHP technologies.

3.8.1.4 Range of Potential Savings

- Grant monies are often available to public utilities for the utilization of digester gas to produce renewable energy. Fuel cells are popular in California because they qualify under the conditions of the Self Generation Incentive Program. Grants typically require a 5-year maintenance service contract for fuel cells.
- Cost savings that might be realized by fuel cell operation are highly dependent on the full cost of comprehensive digester gas treatment, the amount of needed heat, and the local cost of electricity.
- WERF's LCAMER model reports capital costs and O&M costs for phosphoric acid and molten carbonate fuel cells as USD$7,805/kW and USD$0.033/kWh and USD$9,770/kW and USD$0.041/kWh, respectively.

3.8.1.5 Application Potential

- Can be a viable CHP option at wastewater treatment facilities with anaerobic digestion.
- For successful operation it is critical that digester gas, as well as supplemental natural gas, be thoroughly and properly treated in a fail-proof manner prior to delivery to the fuel cell.
- A boiler should be included in the CHP system to cover the fuel cell's heat shortfall to maintain digester temperatures.
- Currently there appears to be only one fuel cell manufacturer actively marketing fuel cells for anaerobic digester gas applications.

The following diagram depicts a typical fuel cell CHP system.

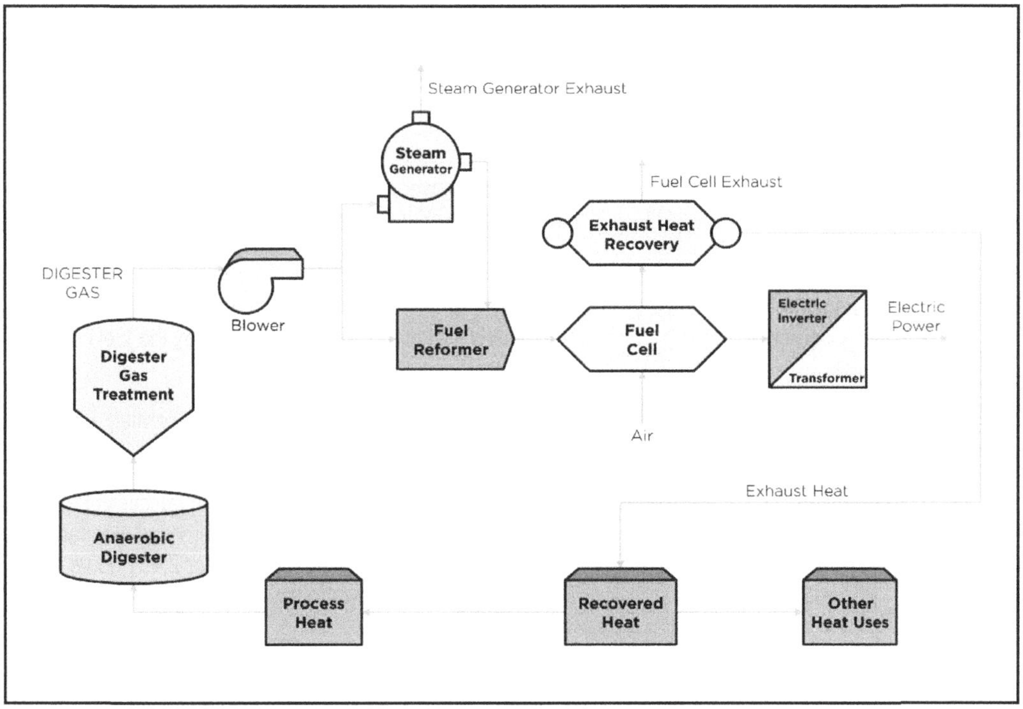

Figure 3-11. Fuel Cells Operated with Anaerobic Digester Gas.
Reprinted with permission from Brown & Caldwell, 2010.

3.8.2 Fuel Cells Operated with Anaerobic Digester Gas South Treatment Plant, King County, Washington, U.S.

Table 3-7. Fuel Cells Operated with Anaerobic Digester Gas Case Study, South Treatment Plant, King County, Washington, U.S..

Ref	Enquiry Item	Response Information, Description and Remarks
1	Location: Country, urban or rural:	Renton, Washington, U.S. Urban.
2	Sector: clean, waste or sludge:	Sludge.
3	Works Owner or Operator: with financial set-up, regulatory or not.	King County.
4	Size: flows and loads or population equivalent:	The South Treatment Plant treats about 435,000 m^3/d of sewage.
5	Energy Provider: with costs, incentives, taxes and conditions:	N/A.
6	Process: physical, chemical, or biological description:	Primary treatment, secondary treatment (activated sludge), and anaerobic sludge digestion.
7	Component: all or part of the works:	Part of the works.
8	Specific energy problem: including quality or consent details:	At full output the fuel cell CHP demonstration project produced about 1 MW of electricity.

Table 3-7. Fuel Cells Operated with Anaerobic Digester Gas Case Study, South Treatment Plant, King County, Washington, U.S., continued.

Ref	Enquiry Item	Response Information, Description and Remarks
9	Process/Plant changes: mechanical, electrical or controls:	Generated power is low voltage direct current and must be inverted to 3-phase alternating current and transformed to 12 kV.
10	Civil/Physical Changes: to water / effluent quality, civil works, or process:	Not aware of any changes to the sewage and biosolids treatment processes that may have been caused by the fuel cell demonstration project.
11	Operational Changes: skill levels, procedures and maintenance routines:	The fuel cell demonstration project was operated and maintained by FueCell Energy.
12	Risks and Dependencies: risk assessment of project and changes.	Fuel quality is a critical consideration for successful fuel cell operation. Due in large part to fuel quality and mechanical issues, the fuel cell operated on digester gas for 2,401 hours during the two-year demonstration project.
13	Implementation: design, build, procurement, installation and commissioning:	The two-year fuel cell demonstration project started in June 2004. King County managed the project and provided some O&M support. The project was funded through annual cooperative agreements from the U.S. EPA Office of Water.
14	Energy Efficiency gains: kWh & kWh/m^3	During the 2,401 hours that the fuel cell operated on digester gas it generated 2.1 million kWh of electrical power, with an electrical efficiency reaching 44%.
15	Cost / Benefit analysis: financial appraisal or payback time.	N/A.
16	Project review: could it be improved or developed?	One of the important outcomes of the demonstration project was that FuelCell Energy was able to gain operational experience which has lead to subsequent fuel cell improvements that are now part of newer generation fuel cells.
17	Confidence grade: on data provided.	A site visit was not conducted to aid in the preparation of this case study. Confidence of the data provided is low.

3.8.2.1 Observations

The data used to generate this case study was obtained at: http://www.kingcounty.gov/environment/wastewater/EnergyRecovery/FuelCellDemonstration/Library/FuelCellExecSum.aspx.

Figure 3-12. Fuel Cell Demonstration Project at the South Treatment Plant, in King County, Washington, U.S..
Source: U.S. EPA.

3.9 Co-Generation – Microturbines

3.9.1 Process Description

- Microturbines are a relatively new CHP technology that are best characterized as much smaller versions of combustion gas turbines.
- Having evolved from large engine turbochargers and similar high-speed turbo machinery, microturbines are essentially small high-speed recuperated combustion gas turbines.
- Microturbines are fully packaged modular machines that comprise the smallest capacity CHP units available.
- Some of the new technologies featured in microturbines include extended-surface recuperators, non-lubricated air bearings, and ultra-fast operating speeds.

3.9.2 Potential Benefits

- Exhaust emissions are amongst the lowest of all CHP prime movers. Air pollutants from microturbines are mainly CO and NOx, with some VOCs, while SOx and particulate matter are negligible.
- Feature dry, low-NOx technology for lean combustion and achieve their lowest emissions when operated at full load.
- Recovered waste heat is available for digester heating or other heating needs in the form of either hot water or low-pressure steam. Similar to the larger combustion gas turbines, heat recovery is only available from the microturbine exhaust.
- Suitable for outdoor installation and does not require a dedicated building for sound attenuation.

3.9.3 Concerns

- Electrical and thermal efficiencies are relatively low compared to other CHP prime movers.
- Require exceptionally clean fuel. Many microturbine installations have been taken out of service because of operational and maintenance issues caused by poor fuel quality.
- A microturbine's recuperator is used to preheat the combustion air with a portion of the exhaust heat. While improving electrical efficiency, recuperators limit the overall heat recovery from microturbines.
- High elevation and warm ambient temperature reduce power generation and fuel efficiency.

3.9.4 Range of Potential Savings

- Complex and expensive gas treatment systems can lower the overall savings potential.
- Grant monies are often available to public utilities for the utilization of digester gas to produce renewable energy.
- WERF's LCAMER model reports capital and O&M costs for digester gas fuelled microturbines as USD$4,124/kW and USD$0.020/kWh, respectively.

3.9.5 Application Potential

- Popularity has increased in recent years because of their clean emissions.
- Offered in relatively small sizes (30-250 kW) making them worth consideration at smaller wastewater treatment facilities with low digester gas volumes.

The figure below depicts a typical microturbine based CHP system.

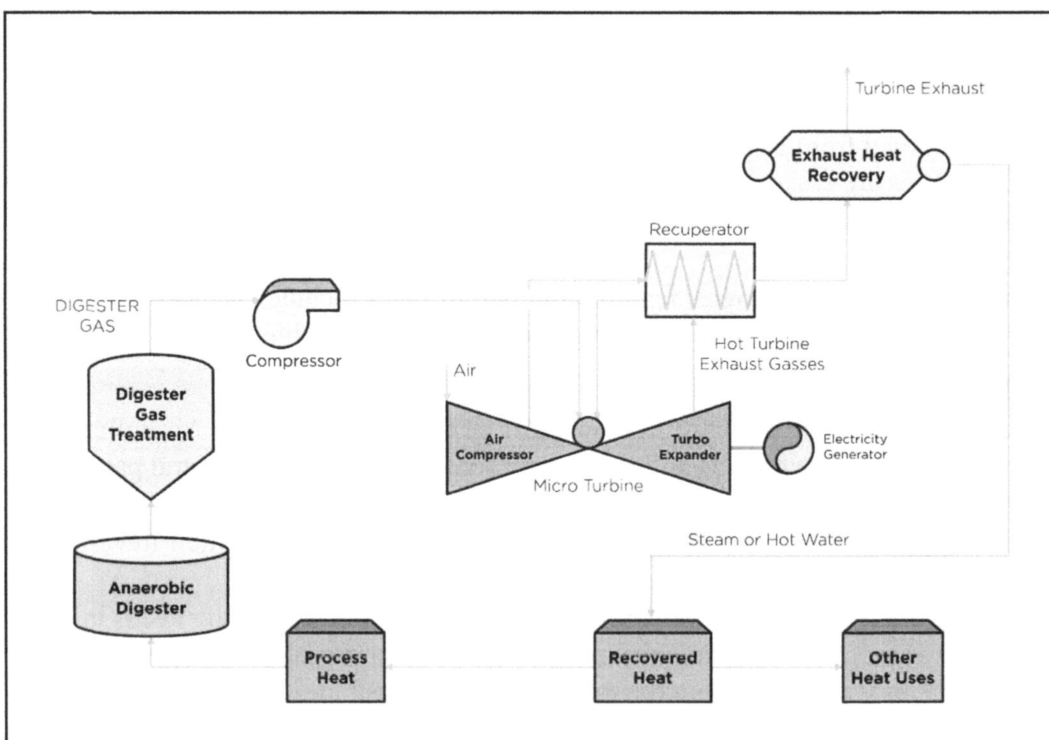

Figure 3-13. Typical Microturbine-based CHP System.
Reprinted with permission from Brown & Caldwell, 2010.

3.9.6 Microturbines Operated with Anaerobic Digester Gas Lancaster Water Reclamation Plant (WRP), Lancaster, CA

Table 3-8. Microturbines Operated with Anaerobic Digester Gas Case Study, Lancaster WRP, Lancaster, California, U.S..

Ref	Enquiry Item	Response information, description and remarks
1	Location: Country, urban or rural:	Los Angeles County, California, U.S. Rural.
2	Sector: clean, waste or sludge:	Sludge.
3	Works Owner or Operator: with financial set-up, regulatory or not.	Los Angeles County Sanitation District.
4	Size: flows and loads or population equivalent:	The Lancaster WRP treats about 57,000 m^3/d of sewage.
5	Energy Provider: with costs, incentives, taxes and conditions:	N/A.
6	Process: physical, chemical, or biological description:	Primary treatment, aeration lagoons, and anaerobic sludge digestion.
7	Component: all or part of the works:	Part of the works.
8	Specific energy problem: including quality or consent details:	The original intention of the microturbine demonstration project was to operate as a true CHP plant with waste heat being used to supplement the existing boiler system for digester heating. The single microturbine utilizes a portion of the WRP's produced digester gas and converts it to about 190 kW of electricity.
9	Process/Plant Changes: mechanical, electrical or controls:	Electrical switchgear synchronizes the renewable energy onto the plant grid. The CHP process also includes digester gas treatment.
10	Civil/Physical Changes: to water / effluent quality, civil works, or process:	The microturbine demonstration has not caused any changes to the sewage and biosolids treatment processes.
11	Operational Changes: skill levels, procedures and maintenance routines:	Microturbine waste heat has periodically exceeded the heat requirements of the anaerobic digesters, causing the boilers to temporarily shut down. Automatically restarting the boilers did not always occur as expected. Because of these challenges the Lancaster WRP continuously operates the digester gas boilers for heating the anaerobic digesters and the microturbine heat recovery system has been taken out of service. Electrical demand at the Lancaster WRP is significantly lower at night than during the day. Night time electrical demand at the WRP is generally less than what the microturbine is capable of producing. So that the microturbine can continue utilizing the available digester gas and operate near its full output, a load bank converts excess electrical power to waste heat.

Table 3-8. Microturbines Operated with Anaerobic Digester Gas Case Study, Lancaster WRP, Lancaster, California, U.S., continued.

Ref	Enquiry Item	Response information, description and remarks
12	Risks and Dependencies: risk assessment of project and changes.	Since start-up the turbine and recuperator have both been replaced twice. Challenges related to the electrical and heat recovery systems have limited the projects performance and availability. The project includes a deep refrigeration system for digester gas treatment. The deep refrigeration system has been very challenging to operate and Ingersoll-Rand reportedly plans to discontinue use of this type of digester gas treatment with its microturbines.
13	Implementation: design, build, procurement, installation and commissioning:	The microturbine demonstration project began operation in March 2005.
14	Energy Efficiency gains: kWh & kWh/m^3	The microturbine generates approximately 190 kW of electricity. Electrical efficiency data is not available.
15	Cost / Benefit analysis: financial appraisal or payback time.	N/A
16	Project review: could it be improved or developed?	The demonstration project has not met the target goal of 90% capacity factor. More operational experience from other microturbine installations is needed to better assess this digester gas utilizing technology.
17	Confidence grade: on data provided.	Very little data was available.

3.9.6.1 Observations

Figure 3-14 Ingersoll-Rand – MT 250 Microturbine at the Lancaster WRP.
Reprinted with permission from Brown & Caldwell, 2010.

3.10 Wind Power

3.10.1 Technology Overview

3.10.1.1 Process Description

- Winds are caused by the uneven heating of the atmosphere by the sun, the irregularities of the earth's surface, and the earth's rotation. Wind flow patterns are modified by the earth's terrain, water bodies, and vegetation. The term wind energy, or wind power, describes the process by which the wind is used to generate mechanical energy or electricity. Wind turbines convert the kinetic energy in the wind into mechanical energy or electrical power.
- Utility-scale turbines range in size from 100 kW to as large as several megawatts (MW). Larger turbines are grouped together into wind farms, which provide bulk power to the electric grid. Single, small turbines (below 100 kW), can be used for low electric load applications such as homes, telecommunication dishes, or water pumping. Small turbines can also be used in connection with diesel generators, batteries, and photovoltaic systems. Such systems are called hybrid wind systems and are typically used in remote, off-grid locations where a connection to the utility grid is not available.
- Wind turbines are categorized into two basic types, depending on the axis orientation of the rotating shaft. Wind turbines having the rotational axis parallel to the earth's surface are known as horizontal-axis type wind turbines, while the wind turbines with the rotational axis perpendicular to the earth's surface are called vertical-axis type wind turbines or Darrieus wind turbines.

Figure 3-15 presents the different components of a horizontal-axis type wind turbine.

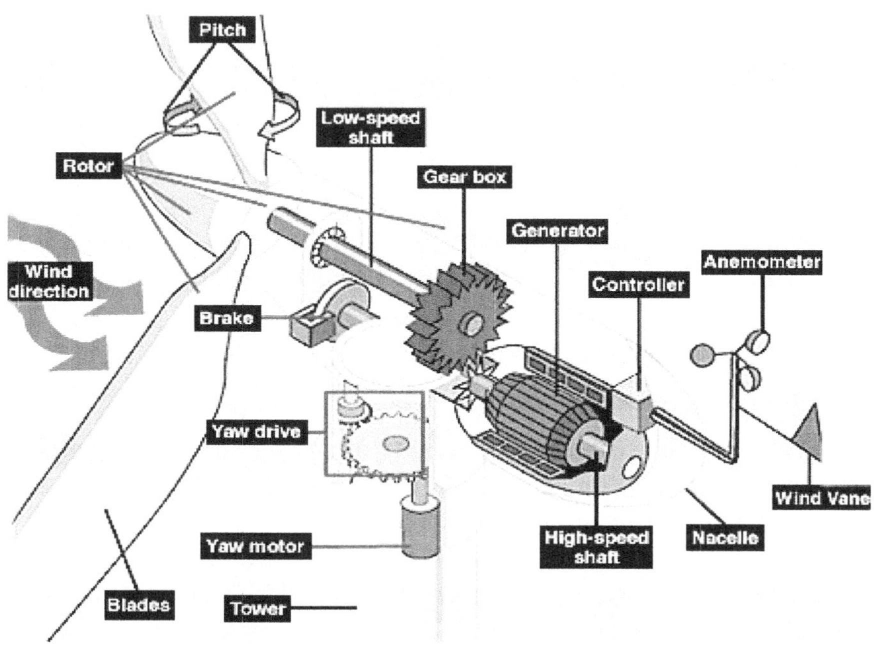

Figure 3-15. Components of a Horizontal – Type Wind Turbine.

Source: www1.eere.energy.gov.

A weather station located on top of a wind turbine turns the nacelle (the enclosure that houses the mechanical and electrical equipment) facing into the wind. The turbine's blades pitch to maximize the speed at which the turbine spins. A shaft attached to the rotating blades rotates a gear box that is attached to an electric generator. AC power produced by the generator is transported to the ground by wires, which are then attached to the electric transmission system.

3.10.1.2 Potential Benefits

- ♦ Wind energy offers several benefits over conventional fossil power, including the following:
  - Wind energy is renewable (inexhaustible) and non-polluting.
  - Wind energy reduces the reliance on non-renewable (exhaustible) fossil fuel power.
  - Wind energy generation is compatible with mixed land use such as agriculture, water treatment, and WWTF operations
  - Wind energy sale is not susceptible to price fluctuations
  - Wind energy projects are modular in nature, and the development and construction of a wind project is relatively fast compared to a new coal or nuclear power plant
  - Wind energy provides water and wastewater utility owners savings in electrical costs and/or a new source of long-term revenue (sale of electric power, leasing of space to a private developer, etc.) without significant impact on existing operations

3.10.1.3 Concerns

- ♦ Contrary to the benefits mentioned above, there are several concerns associated with the development and operation of wind projects that will require mitigation:
  - Impact of wind farms on endangered or protected species
  - Impact of wind turbines on wildlife such as birds and bats. Some wind power generation facilities have instituted pre- and post-construction monitoring programs to assess the impact on wildlife
  - Impact of sound and aesthetics on the local community. Wind turbines can range anywhere from 200 to 400 feet in height. They also can produce a "whoosh" sound in moist air
  - Impact of wind turbines on the local air traffic

3.10.1.4 Range of Potential Cost and Savings

- ♦ Many factors contribute to the cost and productivity of a wind farm. The power that a wind turbine generates is a function of the cube of the average wind speed, which means that small differences in wind speed significantly impact productivity and power generation cost. Power generated by a wind turbine is also a function of the rotor swept area. Since the swept area is a function of the square of the blade length, a modest increase in blade length increases energy capture and cost-effectiveness
- ♦ Utility-scale wind power development projects can cost approximately USD$ 2 million per MW of generating capacity installed. Advantages of economies of scale can be realized by increasing the size of the wind farm. Utilities should consider conducting a feasibility study to assess the wind potential on site, and to develop a planning level capital cost for the project being considered. Financing methods also impact project economics. Costs can be cut

significantly by securing lower-cost financing (state revolving funds, bonds, etc.) that is typically available to utilities or be entering into a joint ownership arrangement (public/private partnership). Furthermore, such projects may qualify for federal and state incentives, which could reduce costs and encourage more favorable investment. As an example, the ARRA of 2009 includes a three year extension of the renewable energy PTC and a new program that allows renewable energy developers the option of foregoing the PTC and securing a grant from the U.S. Treasury Department in the amount of a 30% investment tax credit.

3.10.1.5 Application Potential

- In 2008, the U.S. wind energy industry installed more than 8,500 MW of new generating capacity, increasing the nation's total wind power generating capacity by 50% to over 25,300 MW. Wind power was second only to natural gas in terms of new capacity added. The new wind projects completed in 2008 account for about 42% of the entire new power-producing capacity added nationally last year.
- The impact of wind turbines on water and wastewater utility operations is minimal. Wind power can be produced effectively with minimum operational supervision. Wind turbine technology has made significant progress in recent years. Today's wind turbines are more efficient and cost-effective; however, they are also more complex. Turbine availability (reliability) is a major factor in project success and reducing power generation costs, and the services of a professional familiar with the operation and maintenance of the wind turbines can add significant value. Typically, wind farm owners can execute long-term O&M contracts with wind turbine manufactures for remote, real-time monitoring of the system and periodic onsite maintenance.

3.10.2 Wind Power Case Study: Jersey-Atlantic Wind Farm, Atlantic County Utilities Authority (ACUA) WWTF, Atlantic City, New Jersey, U.S.

Table3-9. Wind Power Case Study: Jersey-Atlantic Wind Farm, ACUA WWTP, Atlantic City, New Jersey, U.S.

Ref	Enquiry Item	Response Information, Description and Remarks
1	Location: Country, urban or rural:	Industrial area within an urban environment.
2	Sector: clean, waste or sludge:	Wastewater
3	Works Owner or Operator: with financial set-up, regulatory or not.	Owner and Operator: ACUA Organization set-up: Regulated public agency Source of revenue: Revenue is realized by charging user fees for wastewater collection and treatment services provided.
4	Size: flows and loads or population equivalent:	Permitted Capacity: 1.75 m^3/s (40 mgd) The monthly average flow to the ACUA WWTF for August 2009 was 1.5 m^3/s (33.4 mgd). The monthly average influent BOD loading to the ACUA WWTF for August 2009 was 24,840 kg/day.

Table 3-9. Wind Power Case Study: Jersey-Atlantic Wind Farm, ACUA WWTF, Atlantic City, New Jersey, U.S., continued.

Ref	Enquiry Item	Response Information, Description and Remarks
5	Energy Provider: with costs, incentives, taxes and conditions:	The daily average electrical demand to operate the ACUA WWTF is 2.5 megawatts (MW). 3 percent of the average annual electrical demand is met with an Authority-owned, onsite 500 kW solar power generation system. 67 percent of the average annual electrical demand is met with wind power produced by the wind farm. A long-term power purchase agreement with the wind farm owner provides wind power at a fixed cost of USD$0.0795 per kWh through the end of 2024.The Authority does not realize any revenue from renewable energy credits or GHG offsets, since the wind farm is owned by a private developer. The WWTFP buys commodity power from Atlantic City Electric to meet the remaining electrical demand. The Authority bids for commodity power every year or two. The 2009 cost for commodity power is USD$0.12 per kWh (USD$0.086 per kWh for power and USD$0.035 per kWh for transmission and distribution).
6	Process: physical, chemical, or biological description:	Liquid Treatment Process: Primary treatment, secondary treatment and disinfection.
7	Component: all or part of the works:	Primary Treatment – bar screens, primary clarification, seepage screening. Secondary Treatment – activated sludge followed by secondary clarification. Disinfection of secondary effluent with sodium hypochlorite prior to discharge. Solids Processing – gravity thickening for primary sludge, centrifuge thickening for WAS, centrifuges for dewatering, and multiple hearth incinerators for incineration of dewatered cake. Incinerator ash disposed of in landfill.
8	Specific energy problem: including quality or consent details:	Rising costs of electricity resulted in increased operational costs. The cost of electricity was the largest component of the WWTF operational costs.
9	Process/Plant changes: mechanical, electrical or controls:	Not applicable
10	Civil/Physical Changes:to water/effluent quality, civil works, or process:	A 7.5 MW wind farm was installed at the ACUA WWTF in 2005. The wind farm consists of five wind turbines, each of which produces 1.5 MW of electric power at wind speeds of 19 km/hr (12 mph) and greater.

Table 3-9. Wind Power Case Study: Jersey-Atlantic Wind Farm, ACUA WWTF, Atlantic City, New Jersey, U.S., continued.

Ref	Enquiry Item	Response Information, Description and Remarks
11	Operational Changes: skill levels, procedures and maintenance routines:	The wind farm is owned and operated by a private firm called the Jersey-Atlantic Wind Farm, LLC. The private firm has entered into a long-term service contract with the wind turbine manufacturer General Electric for periodic inspection and maintenance services. The ACUA operational staff provides support to the manufacturer's maintenance staff as needed during site visits. General Electric uses real-time remote monitoring and control to monitor wind farm performance, and to address operational problems. A large percentage of the problems are resolved by the manufacturer remotely.
12	Risks and Dependencies: risk assessment of project and changes.	The project had to overcome the following challenges: Community acceptance: The Authority had to gain community acceptance for the project because of concerns regarding aesthetics and noise. The project is located in an industrial area within Atlantic City. As a result, aesthetics was not a major concern. The wind turbine operation is practically a silent process; however, in moist air a slight "whoosh" can be heard. Environmental impact of wind turbines: The Authority had to gain acceptance from the Audubon Organization over concerns of impacts of the wind turbines on birds. *Audubon's mission is to conserve and restore natural ecosystems, focusing on birds, other wildlife, and their habitats for the benefit of humanity and the earth's biological diversity.* (Source: http://www.audubon.org/nas/) The Audubon Organization periodically monitors the impact of the wind turbines on birds. The Authority adopted a post-construction monitoring program to assess the impact of the wind farm on birds. Thirty one (31) bird kills have been reported since the wind farm became operational in December 2005. Design Challenges: Since the ACUA WWTF is located in a hurricane zone, the wind turbines and the turbine towers were designed to withstand wind speeds of 217 km/hr (135 mph). Construction challenges: The project developer had to overcome challenges during transportation of large size wind turbine blades and tower components.
13	Implementation: design, build, procurement, installation and commissioning:	The ACUA adopted a Design-Build-Own-Operate Model (public-private partnership) for the project. The ACUA entered into a landlord-tenant relationship with the project developer Community Energy, Inc. (http://www.communityenergyinc.com/).The ACUA rented five circular areas to Community Energy, Inc. to install the wind turbines. Each circular area is 15 m (50 ft.) in diameter. To develop, construct, own, and operate the wind farm, the project developer, Community Energy, Inc., partnered with the ACUA, Babcock and Brown (http://www.babcockbrown.com/) and Central Hudson Energy Group, Inc. (http://www.chenergygroup.com/) to form the Jersey-Atlantic Wind, LLC (http://www.njwind.com/project.html).

Table 3-9. Wind Power Case Study: Jersey-Atlantic Wind Farm, ACUA WWTF, Atlantic City, New Jersey, U.S., continued.

Ref	Enquiry Item	Response Information, Description and Remarks
14	Energy Generation: kWh & kWh/m^3	The ACUA requires approximately 2.5 MW of electric power on an average daily basis to operate the treatment plant. The average power demand equates to 1,800,000 kWh of electrical energy on an average monthly basis. The ACUA is successful to meet 70% of its electrical demand on an annual average basis with renewable energy (solar power and wind power). The WWTP owns and operates a 500 kW DC solar power generation system on-site. The solar power produced by the solar cells is fed into the WWTP electric system. On an annual average basis, 3% of the WWTP's electrical demand is satisfied with solar power. The wind farm provides approximately 67% of electricity required by the WWTP on an annual basis. The ACUA purchases commodity electric power from the local electric utility to meet 30% of its electric demand on an annual average basis.
15	Cost/Benefit analysis: financial appraisal or payback time.	The capital cost for the wind farm project was USD$12,500,000.The ACUA did not contribute funding towards the project capital cost. In 2005, to purchase commodity power from the electric utility the Authority incurred a cost of USD$0.11 per kWh (USD$0.076 per kWh for power and USD$0.035 per kWh for transmission) for electricity consumed. In 2009 the Authority purchases electric power from the electric utility at a cost of approximately USD$0.12 per kWh. The ACUA entered into a long-term power purchase agreement with the wind farm project developer to provide electric power at a fixed price of USD$0.0795 per kWh for a 20-year contract period starting in 2005. The ACUA realizes cost savings by avoiding the cost of purchasing commodity power from the electric utility. The Authority realized cost savings of approximately USD$500,000 per year since the wind farm became operational in December 2005.This equates to cost savings of approximately USD$10,000,000 over the contract term of 20 years. The ACUA also earns revenue from renting space to the project developer for the wind farm. The rental revenue realized by the ACUA is not known. Because the wind farm is owned by a private developer, the Authority does not realize any revenue from renewable energy credits or GHG offsets.

Table 3-9. Wind Power Case Study: Jersey-Atlantic Wind Farm, ACUA WWTF, Atlantic City, New Jersey, U.S., continued.

Ref	Enquiry Item	Response Information, Description and Remarks
16	Project review: could it be improved or developed?	The project and business model can be adopted by other utilities in the water and wastewater industry, provided the utility has the space available to install wind turbine(s) and the required wind potential to make the project cost-effective, and can mitigate the environmental and social impacts of wind power generation. The capital cost (2005 U.S. dollars) for the wind farm was approximately USD$1,670 per Wythe industry-wide average capital cost in 2009 dollars is approximately USD$2,000 per kW for a utility-scale wind power generation facility. The business model (public-private partnership) adopted by the ACUA is a low-risk option for a utility. However, the wind power generation costs can be lowered if a wind project is financed and owned by a utility, provided the utility is willing to manage the risks associated with the development of the project and operation of a wind farm.
17	Confidence grade: on data provided.	The fact sheet presented above is based on information published by the ACUA and communications with ACUA staff. On a scale of 1 through 5, the confidence grade provided to the information presented above is 4.

3.10.2.1 Observations

Project Location

The wind farm is located on the site of the ACUA WWTF in Atlantic City, New Jersey, U.S.A and is called the Jersey-Atlantic Wind Farm. The ACUA WWTF is located very close to the Atlantic Ocean shore line. The wind farm is located in an area classified as "marginal wind resource potential" by the U.S. Department of Energy, National Renewable Energy Laboratory. "Marginal Wind Resource Potential" is a classification given to an area where the wind speed at 50 m (164 ft.) above ground surface elevation is between 3.8 and 4.4 m/s (12.5-14.3 feet per second).

Wind Turbines

The Jersey-Atlantic Wind Farm consists of five (5) wind turbines. The wind power generation capacity of each wind turbine is 1.5 MW. The total wind power generation capacity of the Jersey Atlantic Wind Farm is 7.5 MW.

The height of each wind turbine above ground elevation is 80 m (262 ft.), and each turbine tower has a diameter of 4.25 m (14 ft.) at ground elevation. The blades of the each wind turbine are 36.6 m (120 ft.) in length. Thus, the total distance from the ground surface to the tip of the blade is more than 115 m (380 ft.); approximately the height of a 32-storey building.

The rotational speed for the turbine blades is between 10 and 20 revolutions per minute, and the rotational speed is a function of the prevailing wind speed. Considering the length of the blades, at average wind speeds of 21 to 24 km/hr (13 to 15 mph), the tip of the blade is travelling at 193 km/hr (120 mph).

When the prevailing wind speed exceeds 19 km/hr (12 mph), each wind turbine produces 1.5 MW of electricity. At reduced wind speeds, electricity production decreases. When the wind speed exceeds 72 k/hr (45 mph), the turbines shut down to prevent damage to the turbine mechanism.

3.11 Solar Power

3.11.1 Technology Overview

3.11.1.1 Process Description

- Photovoltaics (PV) is the field of technology related to the application of solar cells for energy by converting solar energy (sunlight) directly into electricity. Some materials exhibit a property known as the photoelectric effect that causes them to absorb photons of light and release electrons. When these free electrons are captured, the electric current that results can be used as electricity. Photovoltaic devices are made of various semiconductor materials, including silicon, cadmium sulfide, cadmium telluride and gallium arsenide in single crystalline, multicrystalline, or amorphous forms.
- Because of growing demand for renewable energy sources, the manufacture of solar cells and photovoltaic arrays has advanced dramatically in recent years. Photovoltaic production is the fastest solar energy growing technology, having doubled every two years since 2002 and increased by 110% in 2008). At the end of 2008, the cumulative global PV installations reached 15,200 MW. Roughly 90% of this generating capacity consists of grid-tied electrical systems.
- A number of solar cells electrically connected to each other and mounted in a support structure or frame is called a photovoltaic module. Modules are designed to supply electricity at a certain voltage, such as a common 12-volt system. Multiple modules can be wired together to form an array.

Modules or arrays can be connected in both series and parallel electrical arrangements to produce any required voltage and current combination. PV modules used in utilities may range in size from a few hundred kW to as large as several MW.

- There are three generic approaches for manufacturing commercial solar cells. The most common approach is to process discrete cells on wafers sawed from either single-crystal or multicrystalline silicon ingots. However, in either case, growing or sawing the ingots is a highly energy-intensive process. A more recent, energy-saving approach is to process discrete cells on silicon wafers cut from multicrystalline ribbons. The third approach involves the deposition of thin layers of non-crystalline-silicon materials on inexpensive substrates. It is the least energy-intensive of the three manufacturing approaches for commercial photovoltaics. This last group of technologies includes amorphous silicon cells deposited on stainless-steel ribbon, cadmium telluride (CdTe) cells deposited on glass, and copper indium gallium diselenide alloy cells deposited on either glass or stainless steel substrates (U.S. Department of Energy).

3.11.1.2 Potential Benefits

- Solar power offers the following benefits over conventional fossil power:
  - Solar power is renewable (inexhaustible) and reduces the reliance on non-renewable (exhaustible) fossil fuel power.
  - Solar power generation has the highest density (global mean of 170 W/m^2) among renewable energies.

- Solar power technology has a low carbon footprint (50 g/kWh vs. 950 g/kWh coal) and does not produce emissions during operation (production of solar panels yield with end-wastes and emissions which are manageable using existing pollution control strategies). One MW solar power offsets approximately 913,000 kg of CO_2. This is equivalent to taking 196 cars off the road for one year.
- Solar power projects are modular in nature and the development and construction of a solar power project is relatively fast compared to a new coal or nuclear power plant.
- Solar power provides water and wastewater utility owners with savings in electrical costs and/or a new source of long-term revenue (sale of electric power, leasing of space to a private developer, etc.) without significant impact on existing operations.
- Solar power project may receive State and Federal financial incentives.
- Grid-connected solar electricity can be used locally, reducing transmission/ distribution losses (transmission losses in the U.S. were approximately 7.2% in 1995).

3.11.1.3 Concerns

- Contrary to the benefits mentioned above, the following concerns associated with the development and operation of solar projects may require mitigation:
  - The efficiency of currently used solar cells is generally around 20%, which requires large land areas for installation. However, experimental high-efficiency cells have an efficiency exceeding 40%, which will significantly reduce the footprint requirement in future.
  - Solar power generation is only feasible in certain areas where sunlight is abundant
  - Solar electricity from conventional silicon-based technologies is not available at night and is less available in cloudy weather conditions. Therefore, a storage or complementary power system is required.
  - Solar cells produce DC which must be converted to AC using a grid tie inventor when used in existing distribution grids. This incurs an energy loss of 4-12%.

3.11.1.4 Range of Potential Cost/Savings

- Factors contributing to the cost and saving of the solar power generation systems include project location, PV system module type, solar panel rating, inverter efficiency, capacity factor, and annual electricity usage and average electricity unit cost. Project costs are reduced for modules with high solar panel ratings (i.e., 305 W), inverter efficiency (i.e., 96%) and capacity factor (i.e., 20%).
- Large, utility-scale solar power development projects (>100 MW) can cost approximately USD$ 2,500 per kW of generating capacity installed, while a small (<1 MW) utility-scale solar power project can cost as high as USD$ 10,000 per kW of generating capacity installed (CH2M HILL, 2009).Advantages of economies of scale can be realized by increasing the size of the solar power farm. Utilities should consider conducting a study to evaluate the feasibility of solar power implementation on site, and to develop a planning level capital cost for the project being considered. Financing methods also impact project economics. Costs can be cut significantly by securing the lower-cost financing (State revolving funds, bonds, etc.) typically available to utilities or by entering into a joint ownership arrangement (public-private partnership). Furthermore, a project may qualify for Federal and State incentives, which could reduce costs and encourage more favorable investment. As an example, the ARRA of 2009 includes a three-year extension of the renewable PTC and a new program that

allows renewable energy developers the option of foregoing the PTC and securing a grant from the U.S. Treasury Department in the amount of a 30% investment tax credit.

3.11.1.5 Application Potential

- Since 2007, when the cumulative installation capacity of PV installations in the U.S. was 1,300 MW (Clean Edge, 2008), the growth of PV installations is estimated at 30% over the next five years.
- The negative impact of solar power projects on water and wastewater utility operations is minimal. Solar power can be produced effectively with minimal operational supervision, and the panels themselves can be built on existing roofs and enclosed infrastructures (i.e., water storage tanks, primary clarifiers). Solar power technology has also advanced dramatically in recent years. One recent advance is a thin film that is lighter and less expensive than the standard crystalline silicone modules. Newer alternatives to the standard modules also include casting wafers instead of sawing, amorphous silicon, microcrystalline silicon. With economies of scale, solar panels become less costly, as manufacturers increase their production to meet increased demand. Existing Federal and State incentives, along with potential carbon credits in the future, may make solar power an attractive solution.
- In some cases, utilities may choose a long-term maintenance contract option with the solar panel manufacturer for monitoring of system performance and periodic onsite maintenance. In other cases, ownership right may belong to either the energy provider or solar system supplier [i.e., Inland Empire Utilities Agency Carbon Canyon Water Reclamation Facility (WRF) Solar Power Generation System], with the utilities allowing use of their land for solar project development.

3.11.2 Solar Power Case Study: Inland Empire Utilities Agency Carbon Canyon WRF, City of Chino, California, U.S.

Table 3-10. Solar Power Case Study:
Inland Empire Utilities Agency Carbon Canyon WRF, City of Chino, California, U.S.

Ref	Enquiry Item	Response Information, Description and Remarks
1	Location: Country, urban or rural	Urban area
2	Sector: clean, waste or sludge	Wastewater
3	Works Owner or Operator: with financial set-up, regulatory or not	Owner and Operator: WWTF is owned and operated by Inland Empire Utilities Agency. Solar Power Generation System is owned and maintained by SunPower Corporation Organization set-up: Regulated public agency Source of revenue: Revenue is realized by charging user fees for wastewater collection and treatment services provided.
4	Size: flows and loads or population equivalent	Permitted Capacity: 43,200 m^3/d (11.4 mgd) Population Equivalent Treatment Capacity: 110,000 capita Average Daily Loads: BOD: 8,900 kg/d TSS: 9,000 kg/d TKN: 1,700 kg/d

Table 3-10. Solar Power Case Study:
Inland Empire Utilities Agency Carbon Canyon WRF, City of Chino, California, U.S., continued

Ref	Enquiry Item	Response Information, Description and Remarks
6	Process: physical, chemical, or biological description	Liquid Treatment Process: Primary treatment, secondary treatment, filtration and disinfection.
7	Component: all or part of the works	Primary Treatment – screening, grit removal, chemically enhanced primary settling, primary effluent flow equalization. Secondary Treatment – activated sludge followed by secondary clarification. Tertiary Treatment – Shallow bed automatic backwash filtration and chlorine disinfection Solids Processing – Solids are pumped to the IEUA's Regional Plant 2 for digestion
8	Specific energy problem: including quality or consent details	Rising costs of electricity resulted in increased operational costs. The cost of electricity was the largest component of the WWTP operational costs. Agency's goal to become 100% self sufficient in producing energy from renewable energy sources to meet 11 MW of electrical energy needed for operating all of its facilities.
9	Process/Plant changes: mechanical, electrical or controls	No changes
10	Civil/Physical Changes: to water/effluent quality, civil works, or process	A 0.7 MW photovoltaic solar system was installed at the WWTP in 2009. The photovoltaic solar system consists of 3,047 solar panels installed on a 1.6-acre area.
11	Operational Changes: skill levels, procedures and maintenance routines	The solar system is owned and maintained by a private solar company called SunPower, which also manufactures the solar panels used at the WWTP. Carbon Canyon Water Reclamation Facility operational staff provides support to the manufacturer's maintenance staff as needed during site visits.
12	Risks and Dependencies: risk assessment of project and changes	The project had to overcome the following challenges: Community acceptance: Information not available. Environmental impact of solar systems: Information not available. However, compared to fossil fuel energy generation system, environmental impacts are minor. Design Challenges: Since Carbon Canyon Water Reclamation Facility WWTP is located in an active earthquake zone, the solar panels were designed and anchored to meet seismic design requirement outlined in ASCE 7. Construction challenges: Minimum
13	Implementation: design, build, procurement, installation and commissioning	The IEUA has adopted a program through which it purchases the power from SunPower Corporation (partnership thru Pacific Gas and Electric), the owner of the solar power system at the facility. Design and installation services were provided separately.

Table 3-10. Solar Power Case Study:
Inland Empire Utilities Agency Carbon Canyon WRF, City of Chino, California, U.S., continued

Ref	Enquiry Item	Response Information, Description and Remarks
14	Energy Generation: kWh & kWh/m^3	The IEUA requires approximately 1.2 MW of electric power on an average daily basis to operate the treatment plant. The average power demand equates to 70,000 kWh to 120,000 kWh of electrical energy on an average monthly basis. The WWTP owns and operates a 0.7 MW DC solar power generation system onsite. The solar power produced by the solar cells is fed into the WWTP electric system, which satisfies 14 % of its electrical demand on an annual average basis. The IEUA purchases commodity electric power from the SunPower (partnership thru Pacific Gas and Electric) to meet its remaining electric demand.
5	Cost/Benefit analysis: financial appraisal or payback time	IEUA, through its power purchase agreement, purchases the electricity from SunPower Corporation (partnership through Pacific Gas and Electric) at a cost of USD$0.11 to USD$0.13 cents per kWh, as per the demand at CCWRF. IEUA is receiving incentives of USD$0.37 per kWh from the California Solar Initiative for the first five years of the project, resulting in an economic savings (net earnings) of approximately USD$0.24 to USD$0.26 per kWh for the first five years of the project. Payback is not applicable to the project (no capital investment).
16	Project review: could it be improved or developed?	The project and business model can be adopted by other utilities in the water and wastewater industry, provided the utility has the space available to install solar panels. The capital cost (2007 USD) for a utility-scale solar power farm is about USD$3,500 per kW. However, recent advances in solar panel material strongly indicate that the average cost will reduce to USD$1,500 per kW in 2020. The business model (public-private partnership) adopted by the IEUA is a low-risk option for a utility. However, the solar power generation costs may be lowered if a solar project is financed and owned by a utility, provided the utility is willing to manage the risks associated with the development of the project and operation of a solar farm.
17	Confidence grade: on data provided.	The fact sheet presented above is based on information published by the IEUA and communications with IEUA staff. On a scale of 1 through 5, the confidence grade provided to the information presented above is 5.

3.11.2.1 Observations

Project Location

The solar system is located on the site of the Inland Empire Utilities Agency (IEUA) Carbon Canyon WRF; City of Chino, California, Sarthe Carbon Canyon WRF is located in industrial urban area. The solar power system is located in an area classified as "good solar resource potential" by the U.S. Department of Energy, National Renewable Energy Laboratory. Figure 3-16 presents picture of solar panels used at the Carbon Canyon WRF.

Figure 3-16. Photograph of the Solar Panel Arrays Used in IEUA Carbon Canyon WRF.
Source: CH2M HILL.

Solar Panels. The Carbon Canyon WRF Solar Power System consists of 3,047 solar panels (SunPower 230-WHT). The total solar power generation capacity of the photovoltaic system is 0.7 MW.

3.12 Hydro Power

3.12.1 Technology Overview

3.12.1.1 Process Description

- Hydroelectricity is electricity generated by hydropower – i.e., the production of power through use of the gravitational force of falling or flowing water. Most hydroelectric power comes from the potential energy of dammed water driving a water turbine or generator. In this case, the energy extracted from the water depends on the volume and on the difference in height between the source and the water's outflow. This height difference is called the head. The amount of potential energy in water is proportional to the head available.
- It is the most widely used form of renewable energy, with an installed capacity of 777 GW supplied 2,998 TWh of hydroelectricity worldwide in 2006. This was approximately 20% of the world's electricity and accounted for about 88% of electricity from the renewable sources.
- While hydroelectric power is the most widely used method in the world, hydroelectric power generation installations in WWTFs are very limited. Example applications include Massachusetts Water Resource Authority Deer Island WWTP in Massachusetts (generates

10% of its energy from a large hydroturbine installed in the ocean outfall) and Point Loma WWTF, San Diego, California (1,350 kW hydroturbine installed in the ocean outfall).

- Once a hydroelectric complex is constructed, the project produces no direct waste, and has a considerably lower output level of the GHG emissions than energy plants powered by fossil fuels.
- A process schematic of a typical hydroelectric generator is presented in Figure 3-17.

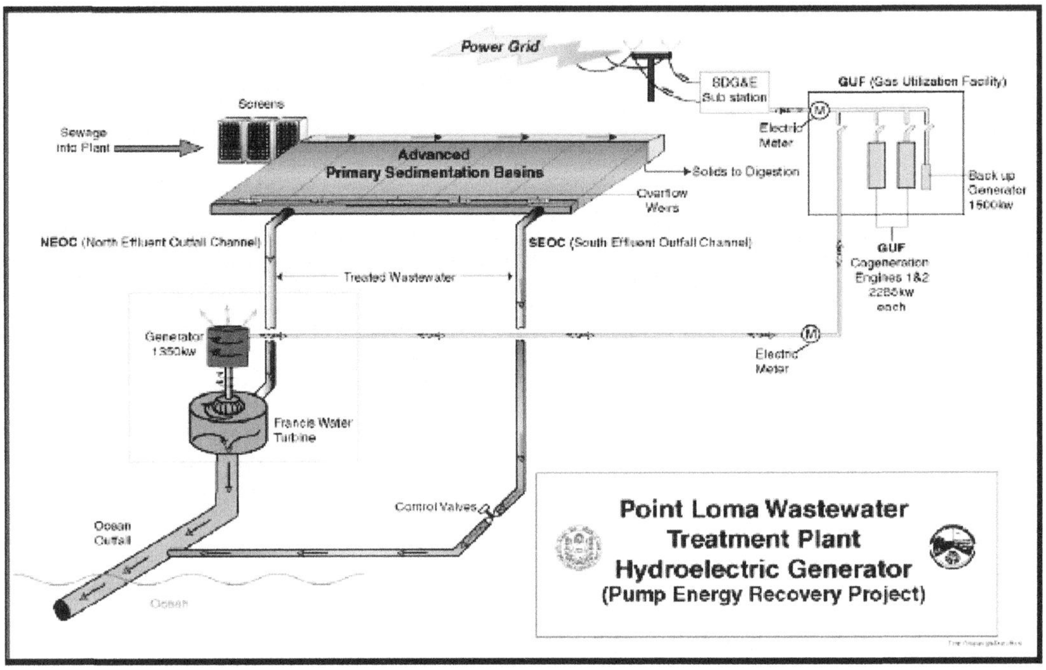

Figure 3-17. Process Schematic of a Typical Hydroelectric Generator.
Reprinted with permission from the City of San Diego.

3.12.1.2 Potential Benefits

- Hydroelectric generation offers several benefits over conventional fossil power, including:
  - Hydroelectric power is renewable (inexhaustible) and reduces the reliance on non-renewable (exhaustible) fossil fuel power.
  - It has the lowest carbon footprint technology among the renewable and non-renewable energy sources (11 g/kWh vs. 900 g/kWh coal).
  - It does not require additional structures to contain equipment and, therefore, does not contribute to the plant footprint.
  - It provides water and wastewater utility owners with savings in electrical costs and/or offers a new source of long-term revenue (sale of electric power, leasing of space to a private developer, etc.) without a significant impact on existing operations.
  - As in other renewable energy options, hydroelectric generation projects may receive State and Federal financial incentives.

3.12.1.3 Concerns

- Contrary to the benefits mentioned above, there are a few concerns associated with the development and operation of hydroelectric generation that may require mitigation, including:
  - Hydroelectric generation is feasible in wastewater facilities where a sufficient head (i.e., 20 m or more) can be maintained between plant and discharge point to create a potential energy.
  - Silt and particulate material in wastewater can cause abrasion on turbine blades.
  - The maintenance requirement is relatively high compared to other renewable energy options (Corrective and scheduled maintenance time may be as high as 20%).

3.12.1.4 Range of Potential Cost/Savings

- Factors that contribute to the cost and savings of the hydroelectric generation systems include project location, availability factor (a factor is applied to account for corrective and scheduled maintenance when the turbine is not available for power generation), and annual electricity usage and average electricity unit cost.
- The cost of a 1,350 kW hydro turbine in 1995 was USD$332,000.Advantages of economies of scale can be realized by increasing the turbine size. Financing methods also impact project economics. Costs can be cut significantly by securing the lower-cost financing (State revolving funds, bonds, etc.) that is typically available to utilities or by entering into a joint ownership arrangement (public-private partnership). Furthermore, a project may qualify for Federal and State incentives that could reduce costs and encourage more favorable investment. As an example, the ARRA of 2009 includes a three-year extension of the renewable energy PTC and a new program that allows renewable energy developers the option of foregoing the PTC and securing a grant from the U.S. Treasury Department in the amount of a 30% investment tax credit.

3.12.1.5 Application Potential

- The cumulative installation capacity of hydroelectric generation installations in the U.S. has reached 79,500 MW, with several hydro turbine suppliers offering varying capacities for hydroelectric projects.
- The impact of hydroelectric projects on wastewater utility operations is minimal. A hydroelectric system can be applied to the wastewater facilities with available head while discharging flow via ocean outfall or similar structures. It can be installed into the existing discharge lines. Existing Federal and State incentives, and potential carbon credits in the future, may make hydroelectric power an attractive solution.

3.12.2 Hydropower Case Study, City of San Diego Point Loma WWTP Hydroelectric Generation System, Point Loma, California, U.S.

Table 3-11. Hydro Power Case Study, City of San Diego Point Loma WWTP Hydroelectric Generation System, Point Loma, California, U.S.

Ref	Enquiry Item	Response information, description and remarks
1	Location: Country, urban or rural	Urban area
2	Sector: clean, waste or sludge	Wastewater
3	Works Owner or Operator: with financial set-up, regulatory or not	Owner and Operator: WWTF is owned and operated by City of San Diego Organization set-up: Regulated public agency Source of revenue: Revenue is realized by charging user fees for wastewater collection, treatment services provided and electricity sell.
4	Size: flows and loads or population equivalent	Permitted Capacity is 908,000 m^3/d (240 mgd) Average Current Flow is 662,000 m^3/d (175 mgd) Population Equivalent of Plant Capacity is 2,400,000 capita Average Daily Loads: BOD: 185,000 kg/d TSS:195,000 kg/d
5	Energy Provider: with costs, incentives, taxes and conditions	The WWTF is energy self-sufficient. No need for purchasing electricity.
6	Process: physical, chemical, or biological description	Liquid Treatment Process: Primary treatment
7	Component: all or part of the works	Primary Treatment – screening, grit removal, chemically enhanced primary settling, final screening prior to ocean discharge Solids Processing – Solids are pumped to the anaerobic digesters and biogas generated is used in two 2.25 MW co-generation facility for electricity generation.
8	Specific energy problem: including quality or consent details	Rising costs of electricity resulted in increased operational costs. The cost of electricity was the largest component of the WWTF operational costs. Agency's goal to become 100% self energy sufficient
9	Process/Plant changes: mechanical, electrical or controls	No changes
10	Civil/Physical Changes: to water/effluent quality, civil works, or process	A 1.3 MW hydroturbine was installed into 3.6 m (12 ft.) discharge line.
11	Operational Changes: skill levels, procedures and maintenance routines	Scheduled maintenance of the hydroturbine by the supplier (once a year) As needed maintenance by WWTF plant operators

Table 3-11. Hydro Power Case Study, City of San Diego Point Loma WWTP Hydroelectric Generation System, Point Loma, California, U.S., continued.

Ref	Enquiry Item	Response information, description and remarks
12	Risks and Dependencies: risk assessment of project and changes	Unknown
13	Implementation: design, build, procurement, installation and commissioning	Not available
14	Energy Generation: kWh & kWh/m^3	The daily average electrical demand to operate the WWTF is approximately 4.5 MW) The average demand is fully satisfied with an onsite co-generation facility (4.53 MW) that utilizes methane gas collected from the anaerobic digesters. A 1,350 kilowatt hydroelectric plant captures the energy of the effluent as it flows down the outfall connection. The power plant, partially funded by a grant from the California Energy Commission, produces up to 1.35 megawatts for sale to the electric grid, enough power to supply energy to 10,000 homes. The City owns and operates other renewable energy sources landfill gas and solar power that make the City self energy-sufficient.
15	Cost/Benefit analysis: financial appraisal or payback time	The capital cost for the investment was USD$332,000 in 1995. Using availability rate of USD$0.82 and electricity sell rate of USD$.03, hydroelectric power generation yielded a payback period of 3.7 years.
16	Project review: could it be improved or developed?	The project and business model can be adopted by other utilities in the wastewater industry, provided the utility has suitable hydraulic conditions to implement the project. The business model adopted by City of San Diego is a fairly low-risk option for the utility.
17	Confidence grade: on data provided	The fact sheet presented above is based on information published by the City of San Diego, Metropolitan Wastewater Department and communications with City of San Diego staff. On a scale of 1 through 5, the confidence grade provided to the information presented above is 5.

3.12.2.1 Observations

Project Location

The Point Loma WWTF is located on a 40-acre site at the western end of Point Loma, California. The treatment plant is located in an urban area. Figure 3-18 presents an aerial photograph of the Point Loma WWTF.

Figure 3-18. Aerial Photograph of the Point Loma WWTF, Located in an Urban Area Point Loma.
Reprinted with permission from the City of San Diego.

3.13 Anaerobic Treatment of Municipal Wastewater – Upflow Anaerobic Sludge Blanket (UASB) Reactors

3.13.1 Technology Overview

3.13.1.1 Process Description

- As most developing countries embark on the costly task of implementing or enhancing wastewater treatment, the issue of considering "appropriate" technologies becomes very relevant. Typically, this consideration is heavily influenced by the construction and the O&M costs (in particular electrical power) of the facility. Given its low capital and operational costs, anaerobic treatment, and UASB technology in particular, is increasingly being considered for municipal wastewater treatment applications in warm-weather locations.
- UASB reactors were developed in the 1970s for the treatment of highly concentrated industrial wastewater. Following the initial reactor designs for the treatment of sugar industry wastes, the benefits of the system – including low sludge production, small footprint, low energy requirements and valuable biogas production – made the UASB reactors an attractive and, hence, widely applicable treatment alternative for highly concentrated industrial wastewaters at mesophilic temperatures. These advantages encouraged investigation of the application of the UASB process to the treatment of domestic wastewater from municipal

applications, where the low BOD concentrations coupled with high particulate BOD fractions results in insufficient CH_4 production for heating the reactor to mesophilic temperatures. Since even moderate wastewater temperatures (i.e., > 18°C) favor satisfactory removal rates within a reasonably sized reactor, the use of low-cost, simple to operate, UASB reactors for domestic wastewater is becoming widespread in tropical countries such as Brazil, Colombia, and India (Foresti, 2002).

- Although this technology cannot by itself produce an advanced effluent of the quality of a conventional secondary process such as activated sludge, it can achieve 60-75% BOD5 removal rates at a fraction of the construction and O&M cost. However, in applications requiring higher treatment levels, UASB reactors are usually followed by a polishing step. Small-scale municipal UASB installations requiring relatively high treatment levels have traditionally relied on simple polishing technologies such as facultative lagoons. This combination has limited applicability in larger installations, as it becomes increasingly difficult to satisfy the space requirements of the lagoons. Activated sludge processes, trickling filters, rotating biological contactors, biological aerated filters, aerated lagoons, polishing ponds, and wetlands, have all been used as a polishing step for the UASB pretreated sewage. The excess sludge from the secondary polishing step is sent to the UASB reactors for digestion and thickening (Sandino et al., 2004; Giraldo et al., 2008).
- UASB facilities for municipal applications are designed on the basis of relatively simple parameters such as hydraulic retention time, water depth, and upflow velocity. The following table presents common design parameters for UASB reactors.

Table 3-12. Design Parameters for UASB Reactors for Municipal Wastewater Treatment Applications.

Parameter	Criteria
Hydraulic Retention Time, hrs	8
Max. Upflow Velocity, m/h (ft/s)	0.6 (5.5×10^{-4})
Water Depth, m (ft)	4.5 (15)
Max Flow/Reactor, l/s (mgd)	70 (1.6)

3.13.1.2 Potential Benefits

- With total plant construction costs similar to a primary treatment plant (including sludge handling facilities), a UASB-based plant is simpler, relies less on mechanical components, and can achieve on average double the organic matter removal rates of conventional primary treatment.
- UASB process will generate substantially lower quantities of sludge, thereby reducing the associated sludge disposal costs.
- Anaerobic treatment offers the potential for the implementation of biogas co-generation schemes.

3.13.1.3 Concerns

- Although this technology cannot by itself produce an effluent of the quality of a conventional secondary process like activated sludge, it can achieve significant organic matter removal rates at a fraction of the construction and O&M cost.
- While more "conventional" aerobic treatment processes (such as activated sludge or trickling filters) can reliably achieve BOD_5 removal rates well in excess of 85% on a monthly (and even weekly) basis, the reported 60-75% BOD_5 removal levels typically associated with UASB facilities should be expected only on the basis of longer averaging periods (e.g., annually). In many instancces, this fact, togcthcr with its comparativcly lowcr trcatmcnt efficiencies, dictates the need to follow the UASB process with reliable and cost-effective polishing steps.
- High levels of gaseous hydrogen sulfide should be expected from the anaerobic treatment of municipal wastewaters with a significant background sulfate concentration. High levels of H_2S could become an operator safety concern, promote corrosion of equipment and facilities, and result in odor complaints from neighbors.
- There are still limitations in combining AST with biological nutrient removal and quantification and handling of diffuse emissions of GHGs requires further research.

3.13.1.4 Range of Potential Cost/Savings

- Based on the recent construction of medium-sized UASB facilities (e.g., serving populations above 200.000), per capita construction costs ranging from USD$30-40/P.E. can be estimated.
- In applications where the typical 65-75% BOD5 removal rates of the UASB process are not sufficient to meet the required discharge criteria, a polishing step must be provided. Applicable polishing technologies include facultative lagoons (the least expensive option if space is available), dissolved air flotation, and submerged aerated filters. Resulting construction costs for a facility relying on UASB reactors followed by a polishing step are in the order of USD$35-60/P.E. These values compare well to the USD$80-100/P.E. range considered typical for a conventional mechanical secondary treatment plant.

3.13.1.5 Application Potential

- UASB technology followed by polishing steps is a treatment option is increasingly being considered for implementation in developing countries throughout the world with warm climates (wastewater temperatures in excess of 18°C, approximately). In these countries, low-cost and low-tech solutions are thought to be more applicable than highly mechanized and complex conventional solutions such as activated sludge.

3.13.2 Anaerobic Treatment of Municipal Wastewater with UASB Reactors Case Study – Rio Frio WWTF, Bucaramanga, Colombia

Table 3-13. Anaerobic Treatment of Municipal Wastewater – UASB Reactors Case Study – Rio Frio WWTF, Bucaramanga, Colombia.

Ref	Enquiry Item	Response Information, Description and Remarks
1	Location: Country, urban or rural	Bucaramanga, Colombia; Urban
2	Sector: clean, waste or sludge	Wastewater
3	Works Owner or Operator: with financial set-up, regulatory or not	Empresa de Pública de Alcantarillado de Santander – EMPAS S.A. ESP; public utility agency
4	Size: flows and loads or population equivalent	2009 average: 750 L/s ;16500 kg/d BOD5; 14800 kg/d TSS; 240,000 inhabitants
5	Energy Provider: with costs, incentives, taxes and conditions	Electrificadora de Santander, is a public utility. Cost of energy: between USD$ 0.11 kWh and 0.18 kWh Basic Cost: USD$ 0.10 plus 10% of contribution (for EPS companies). The source of this electricity is hydroelectric power.
6	Process: physical, chemical, or biological description	Physical (preliminary treatment) followed by biological (anaerobic bioreactors followed by facultative polishing ponds). Sludge from the UASB units is dewatered in sand drying beds.
7	Component: all or part of the works	6 mm fine screening; grit channels; 3 upflow anaerobic sludge blanket (UASB) reactors; 2 facultative polishing ponds.
8	Specific energy problem: including quality or consent details	Low-energy solutions were necessary because of inability of the existing user rate structures to finance more energy-intensive conventional processes.
9	Process/Plant changes: mechanical, electrical or controls	The plant was originally designed as a UASB + polishing pond system to deliver a secondary treatment level effluent (i.e. 30/30 BOD5/TSS). Because of odor releases from ponds and the need to expand capacity in limited site, the plant will replace its ponds with an activated sludge polishing step. Biogas from the UASB reactors will be used for cogeneration, to offset additional power demand from new activated sludge process. New 3 mm fine screening will be added downstream of the existing 6 mm units.
10	Civil/Physical Changes: to water/ effluent quality, civil works, or process	See description of process changes above
11	Operational Changes: skill levels, procedures and maintenance routines	The original UASB +polishing pond configuration, which started in 1990, had minimal O&M requirements. Most of the labor requirements are related to the handling and disposal of treatment residuals such as grit, screenings, and waste sludge.
12	Risks and Dependencies: risk assessment of project and changes	UASB process operates well under fairly narrow hydraulic loading conditions. Excess flows during storm events are normally diverted around the UASB units to prevent the washing of the sludge blanket. The other risk is that by being so inexpensive to operate, the user rate structures will have to increase considerably as the plant expands, and it adopt more energy- and O&M-intensive mechanical components that will replace the ponds.

Table 3-13. Anaerobic Treatment of Municipal Wastewater – UASB Reactors Case Study – Rio Frio WWTF, Bucaramanga, Colombia, continued.

Ref	Enquiry Item	Response Information, Description and Remarks
13	Implementation: design, build, procurement, installation and commissioning	The plant was initially implemented as a conventional design-bid-construct project. Expansion/upgrade could be implemented as design/build (turnkey).
14	Energy Efficiency gains: kWh or kWh/m3 before and after implementation	Currently @ 15,200 kW/hr average
15	Cost/Benefit analysis: financial appraisal or payback time	This plant, which had been producing a secondary treatment level effluent, had an initial construction cost of a primary treatment-only facility (~ U.S. USD$32/P.E.). Operational cost were also very low: in the order of USD$0.003/$m^3$ (less than USD$0.1/1,000 gal.)
16	Project review: could it be improved or developed?	A further review of polishing pond replacement options could be conducted to establish if activated sludge is the most appropriate approach. Trickling filters by themselves or a trickling filter/solids contact configuration should be considered as a potentially simpler and less costly polishing option for the UASB units.
17	Confidence grade: on data provided	The cost savings and process performance documented at this plant are typical of other similar systems also in operation in Latin America.

3.13.2.1 Observations

Project Location

The Rio Frio UASB plant located in the city of Bucaramanga, Colombia, is one of the world's oldest operational large-scale UASB plants treating domestic sewage. First built in 1991 for an equivalent population of 160,000 inhabitants, it was expanded first in 1993 to accommodate 240,000 population equivalents (p.e.) and then in 2001 to 320,000 p.e. with a design flow rate of 750 L/s. The first design did not contemplate odor control structures, and diffuse hydrogen sulfide emissions generated complaints from the community. The surface of the reactor was covered in 2001 and the gas collected and treated to minimize odor impacts. New UASB designs have a covered surface and gas collection and treatment.

The plant has mechanical screening followed by grit removal and three UASB reactors. Effluent from the primary AST reactors is directed to two facultative lagoons of 2.4 hectares each. Biogas is collected and flared. Excess sludge is directed to sand drying beds and disposed of in a landfill after drying. A photograph of the Rio Frio plant is presented in Figure 3-19.

According to recent operational data, the Rio Frio UASB reactors reduce the BOD_5 from an influent concentration of 265 mg/L to an effluent of 60 mg/L. The effluent from the UASB is directed to two facultative lagoons that polish the pretreated water to achieve secondary level treatment of 30/30 mg/L of BOD and TSS for final discharge.

Figure 3-19. UASB Reactors – Rio Frio WWTP, Bucaramanga, Colombia.

REFERENCES

Aly, R.E., R.P. Joshi, R.H. Stark, K.H. Schoenbach, and S.J. Beebe (2001). *"Repair Time of Bacteria after Pulsed Electric Field Application."*IEEE International Conference on Plasma Science, Las Vegas, NV, June 17-22.

American Wind Energy Association (AWEA); www.awea.org.

Atlantic County Utilities Authority (ACUA); www.acua.com.

Atlantic County Utility Authority's(ACUA's) website (www.acua.com).

Audubon Organization; www.audubon.org.

Banaszak, J.E., J.M. Alder, I. Lee, P Parameswaran, B.E. Rittmann, and R. Lopez (2009a). *"Making Waste Biosolids a Sustainable Organic Electron Donor for Denitrification using Focused-Pulsed Technology."* Water Environment Federation Nutrient Removal 2009 Conference Proceedings, Washington, D.C., June 28 - July 1.

Banaszak, J.E., P. Burrowes, and R. Lopez (2009b). *"Focused-Pulsed Treatment of Waste Activated Sludge: A Year in Review."* 5th Canadian Residuals and Biosolids Conference Proceedings, Niagara Falls, ON, September 13-15.

Banaszak, J.E., P. Burrowes, G. Daigger, M.B. Enochs, A.M. Klein, B.E. Rittmann, M. Salerno, and P.R. Schlegel (2007). *"Focused-Pulsed Treatment of Waste Activated Sludge:Application to Waste Biosolids Reduction and Increased Methane Production."* WEFTEC 2007, 80th Annual Water Environment Federation Technical Exhibition and Conference, San Diego, CA, October 13-17.

Banaszak, J.E., P. Burrowes, G. Daigger, M.B. Enochs, A.M. Klein, R. Lopez, B.E. Rittmann, and P.R. Schlegel (2008). *"Focused-Pulsed Treatment of Waste Activated Sludge:Application to Waste Biosolids Reduction And Increased Methane Production."* WEFTEC 2008, 81st Annual Water Environment Federation Technical Exhibition and Conference, Philadelphia, PA, March 30-April 2.

Barjenbruch, M. and O. Kopplow (2003). *"Enzymatic, Mechanical and Thermal Pre-treatment of Surplus Sludge."* Advances in Environmental Research, 7(3): 715-720.

Beveridge, J.R., K. Wall, S.J. MacGregor, J.G. Anderson, and N.J. Rowan (2003). *"Pulsed Electric Field Inactivation of Spoilage Microorganisms in Alcoholic Beverages and the Influence of Pulse Profile."*Digest of Technical Papers-IEEE International Pulsed Power Conference, pp: 1132-1135.

Bougrier, C., A. Battimelli, J.-P. Delgenes, and H. Carrere (2007a). *"Combined Ozone Pretreatment and Anaerobic Digestion for the Reduction of Biological Sludge Production in Wastewater Treatment."*Ozone: Science and Engineering, 29(3): 201-206.

BP Statistical Review of World Energy, June 2009. www.bp.com

Burrowes, P., Hazlewood, P., Garner, S., Deshevy, R., Key, W., Advances in Biogas Treatment – Keeping Up with Demographic Changes, Presented at the 34th Annual Technical Symposium of the Water Environment Association of Ontario, Deerhurst Inn, April 2005

Chernicharo, C.A.L. (1997) Princípios do tratamento biológico de águas residuárias: Reatores anaeróbios. Vol. 5, Ed. DESA – UFMG, Brasil, ISBN 85-7041-130-8

Choi, H., S.-W. Jeong, and Y.-J. Chung (2006*). "Enhanced Anaerobic Gas Production of Waste Activated Sludge Pretreated by Pulse Power Technique."*Bioresource Technology, 97(2): 198-203.

City of San Diego www.sandiego.gov.

De Lucas, A., L. Rodriguez, J. Villaseñsor; and F.J. Fernádez (2005) Denitrification Potential of Industrial Wastewaters.Water Research, 39 (15), 3715.

Email communication with David Miklosi, Vice President, Sterling Biopower.

Energy Information Administration International Statistics Database: www.eia.doe.gov.

Enersave Fluid Mixers; http://www.enersavemixers.com.

EPA. *"EPA Climate Change - Regulatory Initiatives."* U.S.EPA. November 2, 2009. http://www.epa.gov/climatechange/emissions/ghgrulemaking.html (accessed November 2, 2009).

Electric Power Research Institute (EPRI). Quality Energy Efficiency Retrofits for Wastewater Systems. Palo Alto, CA: Electrical Power Research Institute, 1998.

Externalities of Energy: Extension of Accounting Framework and Policy Applications', European Commission (2005). www.externe.info/expolwp6.pdf.

Fang, J., Z. Piao, and X. Zhang (2006). *"Study on High-voltage Pulsed Electric Fields Sterilization Mechanism Experiment."*The Journal of American Science, 2(2): 39-43.

Foresti, E. (2002). Anaerobic treatment of domestic sewage: established technologies and perspectives. Water Science and Technology, Vol. 45 No 10, pp 181 -186.

Giraldo, E., M. Peña, C. Chernicharo, J. Sandino, and A. Noyola. (2007) Anaerobic Sewage Treatment Technology in Latin-America: A Selection of 20 Years of Experiences, Proceedings of the 80th Annual Conference & Exposition of the Water Environment Federation, San Diego, California.

Gossett, J.M. and R.L. Belser (1982*). "Anaerobic Digestion of Waste Activated Sludge."*American Society of Civil Engineers, Journal of the Environmental Engineering Division, 108: 1101-1120.

Gossett, J.M., J.C. Wilson, D.S. Evans, and P.L. McCarty (1978*). "Anaerobic Digestion of Sludge from Chemical Treatment,"* Journal Water Pollution Control Federation, 50: 533-542.

Haug, R.T. (1977). "*Sludge Processing to Optimize Digestibility and Energy Production."* Journal Water Pollution Control Federation, 49(7): 1713-1721.

Haug, R.T., D.C. Stuckey, J.M. Gossett, and P.L. McCarty (1978).*"Effect of Thermal Pretreatment on Digestibility and Dewaterability of Organic Sludges."* Journal Water Pollution Control Federation, 50(1): 73-85.

Jones, Donnie. Navy Electricity and Electronics Training Series. Naval Education and Training Professional Development and Technnology Center, 1998.

Khanal, S.K., D. Grewell, S. Sung, and J. Van Leeuwen (2007). *"Ultrasound Applications in Wastewater Sludge Pretreatment:A Review."* Critical Reviews in Environmental Science and Technology, 37(4): 277-313.

Kim, J., C. Park, T.-H. Kim, M. Lee, S. Kim, S.-W. Kim, and J. Lee (2003). *"Effects of Various Pretreatments for Enhanced Anaerobic Digestion with Waste Activated Sludge."* Journal of Bioscience and Bioengineering, 95(3): 271-275.

Kim, Y.-K., M.-S. Kwak, S.-B. Lee, W.H. Lee, and J.-W. Choi (2002), *"Effects of Pretreatments on Thermophilic Aerobic Digestion."* Journal of Environmental Engineering, 128(8): 755-763.

Koners, U., V. Heinz, D. Knorr, M. Löffler, and W. Schmidt (2006a). *"Effects of Pulsed Electric Field (PEF) Application on Activated Wastewater Treatment Sludge."* 1st Euro-Asian Pulsed Power Conference, Chengdu, China, September 18-22.

Koners, U., S. Toepfl, V. Heinz, P. Camacho, P. Ginestet, and D. Knorr (2004). *"Application of Pulsed Electric Field Treatment for Sludge Reduction in WWTPs."* In: 2nd European Pulsed Power Symposium (EPPS) 2004. Ed.: International Society on Pulsed Power Applications, Shaker, Aachen. p. 47-52.

Koners, U., W. Schmidt, M. Löffler, F. Schwartz, V. Heinz, and D. Knorr (2007). *"Impact of Pulsed Electric Field (PEF) Application on Particle Size of Activated Wastewater Treatment Sludge."* PARTEC 2007, International Congress on Particle Technology, Nuremberg, Germany, March 27th-29th.

Koners, U., W. Schmidt, M. Löffler, V. Heinz, and D. Knorr (2006b). *"The Effect of Implemented Pulsed Electric Field (PEF) Treatment on the Dehydrogenase Activity of Activated Sludge."* WIT Transactions on Ecology and the Environment, 95: 379-388.

Kopplow, O., M. Barjenbruch, and V. Heinz (2004). *"Sludge Pre-treatment with Pulsed Electric Fields."* Water Science and Technology, 49(10): 123-129.

Kunitomo, S. and T. Obo (2003). *"Biofouling Prevention with a Pulsed Discharge."* Digest of Technical Papers-IEEE International Pulsed Power Conference, pp: 1286-1289.

Lafitte-Trouque, S. and C.F. Forster (2002). *"The Use of Ultrasound and Gamma Irradiation as Pre-treatments for the Anaerobic Digestion of Waste Activated Sludge at Mesophilic and Thermophilic Temperatures."* Bioresource Technology, 84(2): 113-118.

Loeffler, M., W. Schmidt, R. Schuhmann, A. Röttering, J. Neumann, J., and C. Dreesen (2001). "*Treatment of Sewage Sludge with Pulsed Electric Fields."* International Conference on Pulsed Power Applications, Gelsenkirchen, Germany, March 27-29.

Madigan, M.T., J.M. Martinko, and J. Parker (2003). Brock Biology of Microorganisms.Upper Saddle River, NJ, Prentice Hall.

Massart, N., J. Doyle, J Jenkins, J. Rowan, and C. Wallis-Lage. (2008). Anaerobic Digestion – Improving Energy Efficiency With Mixing. Proceedings of WEFTEC, Chicago IL, October 2008

Meeting with Dan Hanthorn, Plant Manager, Corvallis WRF.

Metropolitan Wastewater District (MWWD): www.mwwd.gov.

Metropolitan Wastewater District (MWWD): www.mwwd.gov.

Mizuno, A., M. Hayamizu, and T. Tenma (1989). *"Effect of Pulsed Electric Field on Microbes."*EEE International Symposium on Electromagnetic Compatibility, Part 2:539-543.

Mizuno, A., T. Inoue, S. Yamaguchi, K. Sakamoto, T. Saeki, Y. Matsumoto, and K. Minamiyama (1990).*"Inactivation of Viruses using Pulsed High Electric Field."*Conference Record – IAS Annual Meeting (IEEE Industry Applications Society), pp: 713-719.

Mueller, J.A. (2000). *"Pretreatment Processes for the Recycling and Reuse of Sewage Sludge."*Water Science and Technology, 42(9): 167-174.

Muller, J.A. (2001).*"Prospects and Problems of Sludge Pre-treatment Processes."*Water Science and Technology, 44(10): 121-128.

Muller, J.A., A. Winter, and G. Strunkmann (2004). *"Investigation and Assessment of Sludge Pre-treatment Processes."*Water Science and Technology, 49(10): 97-104.

NACWA. 2008 NACWA Financial Survey. Washington, D.C.: National Association of Clean Water Agencies, 2008.

National Renewable Energy Laboratory (NREL); www.nrel.gov.

Owen, W.F. and P.L. McCarty (1979).*"Improving Digester Methane Yield by Heat Treatment."*SAE Preprints, pp: 672-679.

Parkin, G.F. and W.F. Owen (1986). *"Fundamentals of Anaerobic Digestion of Wastewater Sludges."* Journal of Environmental Engineering, 112(5): 867-920.

Phone conversation with Mr. Paul J. Gallagher, Esq., Vice President and General Counsel, ACUA.

Phone conversation with Mr. Tom Alspaugh and Stephanus Suhendra, Senior City Engineers, City of San Diego (09/18/2009).

Rittmann, B.E. and P.L. McCarty (2001).Environmental Biotechnology: Principles and Applications.New York, NY, McGraw-Hill.

Rittmann, B.E., H-S. Lee, H. Zhang, J. Alder, J.E. Banaszak, and R. Lopez (2008). *"Full-scale Application of Focused-Pulsed Pre-treatment for Improving Biosolids Digestion and Conversion to Methane,"* Water Science & Technology, 58(10): 1895-1901.

Rosso, D. and M. Stenstrom (2007). *"Energy-Saving Benefits of Denitrification."* Environemntal Engineer: Applied Research and Practice, 2007: 29-38.

Roxburgh, R., R. Sieger, B. Johnson, B. Rabinowitz, S. Goodwin, G. Crawford, and G. Daigger (2006). *"Sludge Minimization Technologies – Doing More to Get Less."* WEFTEC 2006, 79th Annual Water Environment Federation Technical Exhibition and Conference, Dallas, TX, October 21-25.

Rosso, D. and M. Stenstrom (2007). *"Energy-Saving Benefits of Denitrification."* Environemntal Engineer: Applied Research and Practic, 2007: 29-38.

Science Applications International Corporation (SAIC). Water and Wastewater Energy Best Practice Guidebook. Madison, WI: Focus on Energy, 2006.

Salerno, M., H.-S. Lee, P. Parameswaran, and B.E. Rittmann (2009). *"Using a Pulsed Electric Field as a Pretreatment for Improved Biosolids Digestion and Methanogenesis."* Water Environment Research, 81(8): 831-839.

Sandino J. and C. Yee-Batista. (2000) Evaluation of UASB Effluent Polishing Alternatives for Municipal Applications. Proceedings of the 73rd Conference and Exposition of the Water Environment Federation, Anaheim, CA.

Sandino, J., Rogalla, F., Londoño, J., Sadir R., Sumner, B., Kolarik, J., and Field, A. (2004).Selected Experiences in the Applications of UASB Technology for Large-scale Municipal Advanced Wastewater Treatment Applications, Proceedings of the 77th Annual Conference & Exposition of the Water Environment Federation, New Orleans, Louisiana..

Sato, M. (2007). *"Environmental and Biotechnological Applications of High-voltage Pulsed Discharge in Water."* 28th International Conference on Phenomena in Ionized Gases (ICPIG), Prague, Czech Republic, July 15-20.

Schoenbach, K.H., F.E. Peterkin, S. Beebe, D. Byars, R. Alden, P. Adolphson, and T. Turner (1995). *"Effect of Pulsed Electric Fields on Micro-organisms: Experiments and Applications."* Digest of Technical Papers-IEEE International Pulsed Power Conference, 1: 25-33.

Schoenbach, K.H., R.P. Joshi, R.H. Stark, F.C. Dobbs, and S.J. Beebe (2000). *"Bacterial Decontamination of Liquids with Pulsed Electric Fields."* IEEE Transactions on Dielectrics and Electrical Insulation, 7(5): 637-645.

Schoenbach, K.H., R.W. Alden III, and T.J. Fox (1996*). "Biofouling Prevention with Pulsed Electric Fields."* IEEE Conference Record of Power Modulator Symposium, pp: 75-78.

Schoenbach, K.H.; F.E. Peterkin, R.W. Alden III, and S.J. Beebe (1997*). "Effect of Pulsed Electric Fields on Biological Cells: Experiments and Applications."* IEEE Transactions on Plasma Science, 25(2): 284-292.

Sepulveda, D.R., J.A. Guerrero, and G.V. Barbosa-Canovas (2006). *"Influence of Electric Current Density on the Bactericidal Effectiveness of Pulsed Electric Field Treatments."* Journal of Food Engineering, 76 (4): 656-663.

Speece, R.E. (1996). Anaerobic Biotechnology,Nashville, TN, Archae Press.

Speece, R.E., personal communication, 2007.

Stensel, H.D., O.E. Albertson, and S.E. Strand (2001). *"Status of Waste Sludge Minimization Processes."* World Water and Environmental Resources Congress, Orlando, FL, May 20-24.

Stirling Biopower; http://www.stirlingbiopower.com

Stuckey, D.C. and P.L. McCarty (1978). *"Thermochemical Pretreatment of Nitrogenous Materials to Increase Methane Yield."* Biotechnology and Bioengineering Symposium, 8: 219-233.

Stuckey, D.C. and P.L. McCarty (1984). *"The Effect of Thermal Pretreatment on the Anaerobic Biodegradability and Toxicity of Waste Activated Sludge."* Water Research, 18(11): 1343-1353.

Stumm, W. and J.J. Morgan (1996). Aquatic Chemistry, New York, John Wiley and Sons.

Tchobanoglous, G. and F.L. Burton (1991). Wastewater Engineering: Treatment, Disposal, and Reuse.New York, McGraw Hill.

Timoshkin, I.V., S.J. MacGregor, R.A. Fouracre, B.H. Crichton, and J.G. Anderson (2006). *"Transient Electrical Field across Cellular Membranes: Pulsed Electric Field Treatment of Microbial Cells."*Journal of Physics D: Applied Physics, 39(3): 596-603.

Töpfl, S. (2006). *"Pulsed Electric Fields (PEF) for Permeabilization of Cell Membranes in Food and Bioprocessin – Applications, Process and Equipment Design and Cost Analysis."* Ph.D. Thesis, Department of Food Process Engineering and Food Biotechnology, Berlin University of Technology, Berlin.

Töpfl, S., V. Heinz, and D. Knorr (2007). *"High Intensity Pulsed Electric Fields applied for Food Preservation."* Chemical Engineering and Processing, 46(6): 537-546.

United States Environmental Protection Agency (U.S. EPA) (2003). *"Clean Watersheds Needs Survey 2000."* Office of Wastewater Management Report to Congress: EPA 832-R-03-001, August, 2003.

United States Environmental Protection Agency (U.S. EPA) (2006). "*Emerging Technologies for Biosolids Management."* Office of Wastewater Management Report: EPA 832-R-06-005, September, 2006.

United States Environmental Protection Agency (U.S. EPA) (2008). Ensuring a Sustainable Future: An Energy Managment Guidebook for Wastewater and Water Utilities. Washington, D.C.: U.S. EPA, 2008.

United States Environmental Protection Agency (U.S. EPA); www.epa.gov.

United States Department of Energy (U.S. DOE); www.energy.gov.

Vernhes, M.C., A. Benichou, P. Pernin, P.A. Cabanes, and J. Teissie (2002). *"Elimination of Free-living Amoebae in Fresh Water with Pulsed Electric Fields."*Water Research, 36(14): 3429-3438.

Wallis-Lage, C., J.L. Barnard, and C. DeBarbadillo. Perspectives in Energy Use and Greenhouse Gases at Wastewater Treatment Plants. Proceedings of the 2009 Technical Conference of the Water Environemnt Federation, Orlando, FL.

WEF. Energy Conservation in Water and Wastewater Treatment Facilities. Manual of Practice 32. Alexandria, VA: WEF Press, 2009.

WEF. Design of Municipal Wastewater Treatment Plants, 5th Edition, Manual of Practice 8. Alexandria, VA: WEF Press, 2009.

Wei, Y., R.T. Van Houten, A.R. Borger, D.H. Eikelboom, and Y. Fan (2003). *"Minimization of Excess Sludge Production for Biological Wastewater Treatment."*Water Research, 37: 4453-4467.

WERF. (2008). State of Science Report: Energy and Resource Recovery from Sludge. Alexandria, VA: Water Environment Research Foundation.

Xu, Fei and D. Xiao (2006). *"Analysis of the Effect of Pulsed Electric Field on Cells with the RC Circuit Model and Experiment."* High Voltage Engineering, 32(6): 67-69+83.

Zack, S.I. and G.P. Edwards (1929). *"Gas Production from Sewage Sludge."* Sewage Works Journal, 1: 160.

Zhang, H.Q. (2007). *"Pulsed Electric Fields: Processing System, Microbial and Enzyme Inhibition, and Shelf Life Extension of Foods."* IEEE Transactions on Plasma Science, 35(1): 59-73.

Zhang, Q., A. Monsalve-González, B.L. Qin, G.V. Barbosa-Cánovas, and B.G. Swanson (1994). *"Inactivation of Saccharomyces cerevisiae in Apple Juice by Square-Wave and Exponential-Decay Pulsed Electric Fields."* Journal of Food Process Engineering, 17: 469-478.

Zhang, Q., B.L. Qin, G.V. Barbosa-Cánovas, and B.G. Swanson (1995b). *"Inactivation of Escherichia coli for Food Pasteurization by High-Strength Pulsed Electric Fields."* Journal of Food Processing and Preservation, 19: 103-118.

Zhang, Q., G. V. Barbosa-Cánovas, and B.G. Swanson. (1995a). *"Engineering Aspects of Pulsed Electric Field Pasteurization."* Journal of Food Engineering, 25: 261-281.

WASTEWATER UTILITY

Alabama
Montgomery Water Works & Sanitary Sewer Board

Alaska
Anchorage Water & Wastewater Utility

Arizona
Avondale, City of
Glendale, City of, Utilities Department
Mesa, City of
Peoria, City of
Phoenix Water Services Dept.
Pima County Wastewater Management
Safford, City of
Tempe, City of

Arkansas
Little Rock Wastewater Utility

California
Central Contra Costa Sanitary District
Corona, City of
Crestline Sanitation District
Delta Diablo Sanitation District
Dublin San Ramon Services District
East Bay Dischargers Authority
East Bay Municipal Utility District
El Dorado Irrigation District
Fairfield-Suisun Sewer District
Fresno Department of Public Utilities
Inland Empire Utilities Agency
Irvine Ranch Water District
Las Gallinas Valley Sanitary District
Las Virgenes Municipal Water District
Livermore, City of
Los Angeles, City of
Los Angeles County, Sanitation Districts of
Napa Sanitation District
Novato Sanitary District
Orange County Sanitation District
Palo Alto, City of
Riverside, City of
Sacramento Regional County Sanitation District
San Diego Metropolitan Wastewater Department, City of
San Francisco, City & County of
San Jose, City of
Santa Barbara, City of
Santa Cruz, City of
Santa Rosa, City of
South Bayside System Authority
South Coast Water District
South Orange County Wastewater Authority
South Tahoe Public Utility District
Stege Sanitary District
Sunnyvale, City of
Union Sanitary District
West Valley Sanitation District

Colorado
Aurora, City of
Boulder, City of
Greeley, City of
Littleton/Englewood Water Pollution Control Plant
Metro Wastewater Reclamation District, Denver

Connecticut
Greater New Haven WPCA
Stamford, City of

District of Columbia
District of Columbia Water & Sewer Authority

Florida
Broward, County of
Fort Lauderdale, City of
Jacksonville Electric Authority (JEA)
Miami-Dade Water & Sewer Authority
Orange County Utilities Department
Pinellas, County of
Reedy Creek Improvement District
Seminole County Environmental Services
St. Petersburg, City of
Tallahassee, City of
Toho Water Authority
West Palm Beach, City of

Georgia
Atlanta Department of Watershed Management
Augusta, City of
Clayton County Water Authority
Cobb County Water System
Columbus Water Works
Fulton County
Gwinnett County Department of Public Utilities
Savannah, City of

Hawaii
Honolulu, City & County of

Idaho
Boise, City of

Illinois
Decatur, Sanitary District of
Greater Peoria Sanitary District
Kankakee River Metropolitan Agency
Metropolitan Water Reclamation District of Greater Chicago
Wheaton Sanitary District

Indiana
Jeffersonville, City of

Iowa
Ames, City of
Cedar Rapids Wastewater Facility
Des Moines, City of
Iowa City

Kansas
Johnson County Wastewater
Unified Government of Wyandotte County/ Kansas City, City of

Kentucky
Louisville & Jefferson County Metropolitan Sewer District
Sanitation District No. 1

Louisiana
Sewerage & Water Board of New Orleans

Maine
Bangor, City of
Portland Water District

Maryland
Anne Arundel County Bureau of Utility Operations
Howard County Bureau of Utilities
Washington Suburban Sanitary Commission

Massachusetts
Boston Water & Sewer Commission
Massachusetts Water Resources Authority (MWRA)
Upper Blackstone Water Pollution Abatement District

Michigan
Ann Arbor, City of
Detroit, City of
Holland Board of Public Works
Saginaw, City of
Wayne County Department of Environment
Wyoming, City of

Minnesota
Rochester, City of
Western Lake Superior Sanitary District

Missouri
Independence, City of
Kansas City Missouri Water Services Department
Little Blue Valley Sewer District
Metropolitan St. Louis Sewer District

Nebraska
Lincoln Wastewater & Solid Waste System

Nevada
Henderson, City of
Las Vegas, City of
Reno, City of

New Jersey
Bergen County Utilities Authority
Ocean County Utilities Authority

New York
New York City Department of Environmental Protection

North Carolina
Charlotte/Mecklenburg Utilities
Durham, City of
Metropolitan Sewerage District of Buncombe County
Orange Water & Sewer Authority
University of North Carolina, Chapel Hill

Ohio
Akron, City of
Butler County Department of Environmental Services
Columbus, City of
Metropolitan Sewer District of Greater Cincinnati
Montgomery, County of
Northeast Ohio Regional Sewer District
Summit, County of

Oklahoma
Oklahoma City Water & Wastewater Utility Department
Tulsa, City of

Oregon
Albany, City of
Clean Water Services
Eugene, City of
Gresham, City of
Portland, City of Bureau of Environmental Services
Lake Oswego, City of
Oak Lodge Sanitary District
Water Environment Services

Pennsylvania
Hemlock Municipal Sewer Cooperative (HMSC)
Philadelphia, City of
University Area Joint Authority

South Carolina
Charleston Water System
Mount Pleasant Waterworks & Sewer Commission
Spartanburg Water

Tennessee
Cleveland Utilities
Murfreesboro Water & Sewer Department
Nashville Metro Water Services

Texas
Austin, City of
Dallas Water Utilities
Denton, City of
El Paso Water Utilities

Fort Worth, City of
Houston, City of
San Antonio Water System
Trinity River Authority

Utah

Salt Lake City Corporation

Virginia

Alexandria Sanitation Authority
Arlington, County of
Fairfax, County of
Hampton Roads Sanitation District
Hanover, County of
Henrico, County of
Hopewell Regional Wastewater Treatment Facility
Loudoun Water
Lynchburg Regional Wastewater Treatment Plant
Prince William County Service Authority
Richmond, City of
Rivanna Water & Sewer Authority

Washington

Everett, City of
King County Department of Natural Resources
Seattle Public Utilities
Sunnyside, Port of
Yakima, City of

Wisconsin

Green Bay Metro Sewerage District
Kenosha Water Utility
Madison Metropolitan Sewerage District
Milwaukee Metropolitan Sewerage District
Racine, City of
Sheboygan Regional Wastewater Treatment
Wausau Water Works

Water Services Association of Australia

ACTEW Corporation
Barwon Water
Central Highlands Water
City West Water
Coliban Water Corporation
Cradle Mountain Water
Gippsland Water
Gladstone Area Water Board
Gold Coast Water
Gosford City Council
Hunter Water Corporation
Logan Water
Melbourne Water
Moreton Bay Water
Onstream
Power & Water Corporation
Queensland Urban Utilities
SEQ Water
South Australia Water Corporation
Sunshine Coast Water
Sydney Catchment Authority
Sydney Water
Unity Water
Wannon Regional Water Corporation
Watercare Services Limited (NZ)
Water Corporation
Western Water
Yarra Valley Water

Canada

Edmonton, City of/Edmonton Waste Management Centre of Excellence
Lethbridge, City of
Regina, City of, Saskatchewan
Toronto, City of, Ontario
Winnipeg, City of, Manitoba

STORMWATER UTILITY

California

Fresno Metropolitan Flood Control District
Los Angeles, City of, Department of Public Works
Monterey, City of
San Francisco, City & County of
Santa Rosa, City of
Sunnyvale, City of

Colorado

Aurora, City of
Boulder, City of

Florida

Orlando, City of

Iowa

Cedar Rapids Wastewater Facility
Des Moines, City of

Kansas

Lenexa, City of
Overland Park, City of

Kentucky

Louisville & Jefferson County Metropolitan Sewer District

Maine

Portland Water District

North Carolina

Charlotte, City of, Stormwater Services

Pennsylvania

Philadelphia, City of

Tennessee

Chattanooga Stormwater Management

Texas

Harris County Flood Control District, Texas

Washington

Bellevue Utilities Department
Seattle Public Utilities

STATE

Connecticut Department of Environmental Protection
Kansas Department of Health & Environment
New England Interstate Water Pollution Control Commission (NEIWPCC)
Ohio Environmental Protection Agency
Ohio River Valley Sanitation Commission
Urban Drainage & Flood Control District, CO

CORPORATE

ADS LLC
Advanced Data Mining International
AECOM
Alan Plummer & Associates
Alpine Technology Inc.
Aqua-Aerobic Systems Inc.
Aquateam–Norwegian Water Technology Centre A/S
ARCADIS
Associated Engineering
Bernardin Lochmueller & Associates
Black & Veatch
Blue Water Technologies, Inc.
Brown & Caldwell
Burgess & Niple, Ltd.
Burns & McDonnell
CABE Associates Inc.
The Cadmus Group
Camp Dresser & McKee Inc.
Carollo Engineers Inc.
Carpenter Environmental Associates Inc.
CET Engineering Services
CH2M HILL
CRA Infrastructure & Engineering
CONTECH Stormwater Solutions
D&B/Guarino Engineers, LLC
Damon S. Williams Associates, LLC
Ecovation
EMA Inc.
Environmental Operating Solutions, Inc.
Environ International Corporation
Fay, Spofford, & Thorndike Inc.
Freese & Nichols, Inc.
ftn Associates Inc.
Gannett Fleming Inc.
Garden & Associates, Ltd.
Geosyntec Consultants
GHD Inc.
Global Water Associates
Greeley and Hansen LLC
Hazen & Sawyer, P.C.
HDR Engineering Inc.
HNTB Corporation
Hydromantis Inc.
HydroQual Inc.
Infilco Degremont Inc.
Jason Consultants LLC Inc.
Jordan, Jones, & Goulding Inc.
KCI Technologies Inc.
Kelly & Weaver, P.C.
Kennedy/Jenks Consultants
Larry Walker Associates
LimnoTech Inc.
Lombardo Associates, Inc.
The Low Impact Development Center Inc.
Malcolm Pirnie Inc.
Material Matters, Inc.
McKim & Creed
MWH
NTL Alaska, Inc.
O'Brien & Gere Engineers Inc.
Odor & Corrosion Technology Consultants Inc.
Parametrix Inc.
Parsons
Post, Buckley, Schuh & Jernigan
Praxair, Inc.
RMC Water & Environment
Ross & Associates Ltd.
SAIC
Siemens Water Technologies
The Soap & Detergent Association
Smith & Loveless, Inc.
Southeast Environmental Engineering, LLC
Stone Environmental Inc.
Stratus Consulting Inc.
Synagro Technologies Inc.
Tetra Tech Inc.
Trojan Technologies Inc.
Trussell Technologies, Inc.
URS Corporation
Wallingford Software
Westin Engineering Inc.
Wright Water Engineers
Zoeller Pump Company

INDUSTRY

American Electric Power
American Water
Anglian Water Services, Ltd.
Chevron Energy Technology
The Coca-Cola Company
Dow Chemical Company
DuPont Company
Eastman Chemical Company
Eli Lilly & Company
InsinkErator
Johnson & Johnson
Merck & Company Inc.
Procter & Gamble Company
Suez Environnment
United Utilities North West (UUNW)
United Water Services LLC
Veolia Water North America

Note: List as of 4/20/10

IWA Publishing's authorised EU representative for General Product Safety Regulations is Diane D'Arras, 15 rue Duret, 75116 Paris, France, e-mail: safety@iwap.co.uk.

Printed and bound by CPI Group (UK) Ltd, Croydon, CR0 4YY
17/09/2026
02218810-0013